Unlocking Environmental Narratives

Towards Understanding Human Environment Interactions through Computational Text Analysis

Ross S. Purves, Olga Koblet and Benjamin Adams

]u[

ubiquity press
London

Published by
Ubiquity Press Ltd.
Unit 322–323
Whitechapel Technology Centre
75 Whitechapel Road
London E1 1DU
www.ubiquitypress.com

First published 2022

Cover design by Amber Dalgleish
Cover image by Ross S. Purves
Cover design and image licensed under CC BY 4.0
Print and digital versions typeset by Diacritech Technologies Pvt Ltd.

ISBN (Paperback): 978-1-911529-56-9
ISBN (PDF): 978-1-911529-57-6
ISBN (EPUB): 978-1-911529-58-3
ISBN (Mobi): 978-1-911529-59-0

DOI: https://doi.org/10.5334/bcs

The full text of this book has been peer-reviewed to ensure high academic standards. For full review policies, see http://www.ubiquitypress.com/

Suggested citation:
Purves, R.S., Koblet, O. and Adams, B. 2022. *Unlocking Environmental Narratives: Towards Understanding Human Environment Interactions through Computational Text Analysis*. London: Ubiquity Press. DOI: https://doi.org/10.5334/bcs. License: CC-BY 4.0

To read the free, open access version of this book online, visit https://doi.org/10.5334/bcs or scan this QR code with your mobile device:

Contents

Chapter 8: The Wild Process: Constructing Multi-Scalar Environmental Narratives   **161**
Joanna E. Taylor and Benjamin Adams

Chapter 9: Inferring Value: A Multiscalar Analysis of Landscape Character Assessments   **179**
Joanna E. Taylor, Meladel Mistica, Graham Fairclough and Timothy Baldwin

Chapter 10: Interpreting Natural Spatial Language in a Fictional Text: Analysing Natural and Urban Landscapes in Mary Shelley's *Frankenstein*   **197**
Tobias Zuerrer

Chapter 11: Discovering Spatial Referencing Strategies in Environmental Narratives 211
Simon Scheider, Ludovic Moncla and Gabriel Viehhauser

Chapter 12: Surveying the Terrain and Looking Forward 233
Ross S. Purves, Olga Koblet and Benjamin Adams

List of Figures

List of Tables

Notes about the book

About the cover

The cover image tells a story - it shows glacial striations - scratches left on bedrock by a glacier that has now receded. Such marks can be found in many locations across the world, but this site - the Rhone Glacier in Switzerland is special - since it has been visited by tourists, and written about, for more than 150 years. If we were to turn 180 degrees, then we would see the rapidly receding remains of the glacier where tourists, after paying an entry fee, can descend into ice grottos laboriously carved out of the ice each spring. We would also see thermal blankets, draped across the glacier in an attempt to slow its melt and preserve the grottos for a few more years. This narrative can be pieced together by reading newspaper stories, travel diaries and even fictional writing about this site, and analysing such material at scale, using computers, is the subject of this book.

Code related to this book

We have included the code and public domain data used in the introductory and case study chapters of this book in the following GitHub repository: https://github.com/Unlocking-Environmental-Narratives/Unlocking-Environmental-Narratives-Book, DOI: 10.5281/zenodo.6766825.

Permission

Contributors

Benjamin Adams
Department of Computer Science and Software Engineering, University of
Canterbury, Christchurch, New Zealand
Email: benjamin.adams@canterbury.ac.nz
Web: https://www.canterbury.ac.nz/engineering/contact-us/people/
ben-adams.html
ORCID: https://orcid.org/0000-0002-1657-9809

Benjamin Adams is an Associate Professor of Computer Science at the
University of Canterbury in New Zealand. His expertise lies in the development and application of information retrieval and machine learning algorithms
that aid the collection, management, analysis and use of large spatial data sets
to better support decision-making in complex environments. A key driver
of his research is the goal of developing computational methods that can
translate unstructured and semi-structured data that are created initially for
human communication, such as crowdsourced social media and other natural
language text and images, into forms that aid geographic problem-solving.

Timothy Baldwin[1,2]
[1]Department of Natural Language Processing, The Mohamed bin Zayed University of Artificial Intelligence
[2]School of Computing and Information Systems, The University of Melbourne,
Australia

Timothy Baldwin is Associate Provost Academic and Student Affairs and Act-
ing Head of the Department of Natural Language Processing, The Mohamed
bin Zayed University of Artificial Intelligence in addition to being a Melbourne
Laureate Professor in the School of Computing and Information Systems, The
University of Melbourne. His primary research focus is on natural language
processing, including social media analytics, deep learning and computational
social science.

Ricardo Campos[1,2]

[1]Ci2 - Smart Cities Research Center - ICT Departmental UNIT, Polytechnic Institute of Tomar, Portugal
[2]INESC TEC., Portugal
Email: ricardo.campos@ipt.pt
Web: http://www.ccc.ipt.pt/ ricardo/
ORCID: https://orcid.org/0000-0002-8767-8126

Ricardo Campos is a Professor at the Polytechnic Institute of Tomar and a Senior Research at INESC TEC with more than 10 years of experience in Information Retrieval (IR) and Natural Language Processing (NLP). He is the leading author of the highly impactful Yake! keyword extractor toolkit. His current research focuses on developing methods concerned with the process of narrative extraction from texts. He is particularly interested in practical approaches regarding the relationship behind entities, events and temporal aspects, as a means to make sense of unstructured data.

Graham Fairclough
McCord Centre for Landscape, Newcastle University, UK
Email: graham.fairclough@newcastle.ac.uk
Web: https://www.ncl.ac.uk/hca/people/profile/grahamfairclough.html

Graham Fairclough is an archaeologist and historian with research interests in heritage and landscape studies. Until 2012 he worked in the UK government agency English Heritage, and is presently a research member of the McCord Centre for Landscape at Newcastle University (UK). He has co-edited the interdisciplinary journal Landscapes since 2011 and has published widely on heritage, landscape and sustainability. He coordinated the EU CHeriScape network (2012–2018) and is currently a member of the EU MSC HERILAND project (2018–2022) and the Anglo-German DFG/AHRC project Devastation & Relocation.

Karen Jones
School of History, University of Kent, UK
Email: k.r.jones@kent.ac.uk
Web: https://www.kent.ac.uk/history/people/400/jones-karen
ORCID: https://orcid.org/0000-0002-4787-6830
Twitter: @drkarenjones

Professor Karen Jones is an environmental historian at the University of Kent with a particular interest in place, storytelling and environmental change over time. She has written extensively on parks and conservation landscapes and on species entanglements between humans and other animals, especially wolves. Her current research focuses on city parks, health and urban metabolism, with a particular focus on recovering the voice of marginal communities across a global geography of urban greening.

Olga Koblet
Department of Geography, University of Zurich, Switzerland
E-mail: olga.koblet@geo.uzh.ch
ORCID: http://orcid.org/0000-0002-4298-1789

Olga Koblet is a former postdoctoral researcher at the Geocomputation Group of the University of Zurich. Her focus lies in the use of unstructured text to better understand landscape and landscape perception. In particular, she has worked on new approaches to extracting descriptions of perception from unstructured text and linking these to landscape character assessment.

Katrín Anna Lund
Institute of Life and Environmental Sciences, University of Iceland
E-mail: kl@hi.is
Web: https://www.hi.is/starfsfolk/kl
ORCID: https://orcid.org/0000-0003-1151-5328

Katrín Anna Lund is Professor of Geography and Tourism, Institute of Life and Environmental Sciences at the University of Iceland. She is a social anthropologist with a PhD from the University of Manchester. She has done research in Spain, Scotland and Iceland, as well as in Northern Norway and Finnish Lapland. Her research has focused on tourism, travel and the perception of landscape, but landscape studies have been central to her work on travel and tourism, with a special emphasis on walking and narratives. She has been studying destination making in tourism for the past 10 years.

Sarah Luria
Department of English
College of the Holy Cross, Worcester, Massachusetts
E-mail: sluria@holycross.edu

Sarah Luria is a professor of American Literature and former Director of Environmental Studies at the College of the Holy Cross. She studies how literature and stories more broadly impact a sense of place and relationship to the environment. Her current book project, *The Story of My Street*, gives an in-depth tour of the environmental history of the street where she lives. She recently directed and edited a film on the Native American history and ongoing identity of her college campus and surrounding community.

Diana Maynard
Dept of Computer Science, University of Sheffield, UK
Email: d.maynard@sheffield.ac.uk
Web: https://www.sheffield.ac.uk/dcs/people/research-staff/diana-maynard
ORCID: https://orcid.org/0000-0002-1773-7020
Twitter: @dianamaynard

Diana Maynard is a Senior Research Fellow at the University of Sheffield with more than 30 years of experience in Natural Language Processing (NLP). Since 2000 she has been one of the key developers of the GATE NLP toolkit, leading work on Sheffields open-source multilingual text analysis tools. Her main research interests are in practical, multidisciplinary approaches to text and social media analysis, in a wide range of fields including sustainability and the environment, politics, climate change, natural disasters, human rights, law and journalism.

Meladel Mistica
Melbourne Data Analytics Platform, University of Melbourne, Australia
Email: misticam@unimelb.edu.au
Web: https://findanexpert.unimelb.edu.au/profile/3575-mel-mistica
ORCID: https://orcid.org/0000-0001-6147-120X
Twitter: @undergoer

Meladel Mistica is a researcher at the Melbourne Data Analytics Platform (MDAP), University of Melbourne. She has extensive industry and academic experience in natural language processing (NLP). Her role at MDAP is focused on enabling and guiding academic colleagues from Humanities, Arts and Social Sciences in adopting computational and digital methods in their research.

Ludovic Moncla
Univ Lyon, INSA Lyon, CNRS, UCBL, LIRIS, UMR5205, F-69622, France
E-mail: ludovic.moncla@insa-lyon.fr
Web: https://ludovicmoncla.github.io
ORCID: https://orcid.org/0000-0002-1590-9546
Twitter: @MonclaLudovic

Ludovic Moncla is associate professor in Computer Science at INSA Lyon and at the LIRIS laboratory. His research focuses on multidisciplinary aspects of Natural Language Processing, Information Retrieval, Data Mining, Digital Humanities and Geographic Information Science. He develops methods for geoparsing and geocoding texts in order to solve digital and societal issues related to the territory and the exploitation of textual data.

Ross Purves[1,2]
[1]Department of Geography, University of Zurich, Switzerland
[2]URPP Language and Space, University of Zurich, Switzerland
E-mail: ross.purves@geo.uzh.ch
Web: https://www.geo.uzh.ch/ rsp
ORCID: https://orcid.org/0000-0002-9878-9243
Twitter: @GCUZH

Ross Purves is a professor of Geographic Information Science at the Department of Geography, University of Zurich. His research aims to address societally relevant research questions, with the fundamental aim of making theoretical, thematic and methodological contributions to Geographic Information Science. He is particularly interested in developing methods and answering questions through the use of unstructured information, often in the form of text.

Simon Scheider
Department of Human Geography and Spatial Planning, Utrecht University, the Netherlands
E-mail: s.scheider@uu.nl
Web: https://www.uu.nl/medewerkers/SScheider
ORCID: https://orcid.org/0000-0002-2267-4810

Simon Scheider is an assistant professor in Geographic Information Science at the Department of Human Geography and Spatial Planning, Utrecht University. His research lies at the interface between conceptual modeling, geographic data analysis and knowledge extraction. He is particularly interested in understanding the concepts underlying spatio-temporal data, including reference systems, fields, places, spatial objects, events, trajectories and their relationship to human activities, and in practical methods for modelling and handling them with GIS, Artificial Intelligence and semantic technology.

Joanna E. Taylor
Department of English, American Studies, and Creative Writing, University of Manchester, UK
E-mail: joanna.taylor@manchester.ac.uk
Web: https://www.research.manchester.ac.uk/portal/joanna.taylor.html
ORCID: https://orcid.org/0000-0001-8597-0097
Twitter: @JoTayl0r0

Joanna E. Taylor is Presidential Fellow in Digital Humanities at the University of Manchester. Her research focuses on literary geographies of the long 19th century, spatial poetics and environmental humanities. Digital methodologies and technologies – particularly from Natural Language Processing and Geographical Information Science – extend the reach of this work into investigations into the relationships between a place's cultures, ecologies and heritage.

Gabriel Viehhauser
Department of Digital Humanities, University of Stuttgart, Germany
E-mail: viehhauser@ilw.uni-stuttgart.de
Web: https://www.ilw.uni-stuttgart.de/institut/team/Viehhauser-00001/
ORCID: https://orcid.org/0000-0001-6372-0337

Gabriel Viehhauser is a professor for Digital Humanities at the University of Stuttgart. With a background in German medieval literary studies, his research focuses on digital editions and digital text analysis as well as on a combination of both fields. He is particularly interested in digital narratology and the modelling of space in narratives.

Flurina Wartmann
Department of Geography & Environment, University of Aberdeen, UK
Email: flurina.wartmann@abdn.ac.uk
Web: https://www.abdn.ac.uk/geosciences/people/profiles/flurina.wartmann
ORCID: https://orcid.org/0000-0003-4788-2963
Twitter: @DrFlurina

Flurina Wartmann is a Lecturer in Geography at the University of Aberdeen. Her research interests revolve around the relationship between society and nature, including nature conservation, landscape restoration and rewilding, landscape assessment, cultural landscape values and sense of place. In her research, she uses interdisciplinary approaches and a variety of data sources, including interviews, public surveys, social media photographs, digitised text sources and participatory Geographic Information Systems (GIS).

Tobias Zürrer
Email: tobias.zuerrer@ksa.sz.ch

Tobias Zürrer is a former English literature and geography student at the University of Zurich. He is interested in the conceptualisation of natural and urban landscapes in (historical) fictional texts.

Funding Acknowledgements
Benjamin Adams' work on this book was partially supported by the Strategic Science Investment Fund from Government Funding, administered by the Ministry of Business Innovation and Employment, Aotearoa/New Zealand, through the grant 'Beyond Prediction: explanatory and transparent data science' (UOAX1932).

Ricardo Campos was financed by the ERDF European Regional Development Fund through the North Portugal Regional Operational Programme (NORTE 2020), under the PORTUGAL 2020 and by National Funds through the Portuguese funding agency, FCT - Fundação para a Ciência e a Tecnologia within project PTDC/CCI-COM/31857/2017 (NORTE-01-0145-FEDER-03185). This funding fits under the research line of the Text2Story project.

Sarah Lurias Research was partially funded by a Course Development Grant through the College of the Holy Cross, Worcester, Massachusetts.

Ross Purves gratefully acknowledges funding from the Swiss National Science Foundation (200020E_186389) and the University of Zurich.

Simon Scheider's work was partially funded by the European Research Council (ERC) under the European Union's Horizon 2020 research and innovation programme (grant agreement No. 803498).

Foreword

This book started with an idea: we wanted to bring together a group of people, to work together on a theme that we considered exciting. All three of us are excited by the possibilities of analysing text as a tool for better understanding the world around us, and we were convinced that if we could bring the right people together, in the right place, we could do something worthwhile.

That idea turned into a workshop, with the contributors who have brought this book alive. In spring 2019, we travelled together to a beautiful mountain village in Stels, Switzerland, and spent four wonderful days brainstorming and discussing many of the ideas that came together to form this book. We left the workshop excited by what we had achieved, and with a concrete plan for a set of case studies to populate our original idea. We had a plan, and a timetable, and were confident that we could quickly bring this book to fruition. Of course, as academics, everything (almost) always takes longer than we planned. This book is no exception, and the events that we have all lived through since spring 2019 are part of the narrative of our lives. We have all come to appreciate our surroundings in ways we did not before, and many of us have not seen friends and family for a long time.

Throughout that time, this book has remained a dim flame, sometimes sputtering, and perhaps nearly going out. Occasionally, most often when we received a contribution, the flames got brighter, and we could once more see the end. Writing this material, and more importantly the reading that this work has involved, has once again convinced us of the power, the depth and the beauty of the written word. Of course, words alone do not suffice, and we are especially grateful to Laura Wysling for giving the book a consistent graphical look. All of the contributors have answered our many mails, dealt with our queries and helped us to get this project over the line, and we could not have done it without them. Thanks too are due to the reviewers of our book, and the editorial team at Ubiquity for constructive feedback along the way. And finally, thanks to Franziska Komossa for taking one last look at the text for us.

But the biggest thanks go to our families – over the last three years they have been the people with whom we have spent the most time, who've been there when we needed support, and who have reminded us of the other stuff in life that is important. Thanks.

Ross, Olga and Ben
April, 2022

PART I

Themes, Resources and Methods

CHAPTER 1

Introduction

Ross S. Purves

Department of Geography; URPP Language and Space, University of Zurich, Switzerland

Olga Koblet

Department of Geography, University of Zurich, Switzerland

Benjamin Adams

Department of Computer Science and Software Engineering, University of Canterbury, Christchurch, New Zealand

The beauty or excellence of the Scottish landscape, arises principally from the diversity of surface, its lakes and rivers; its defects arise from its sterility, its want of wood, and perhaps in no inconsiderable degree from the vitiated taste of great proprietors in improving their grounds.

Scots Magazine, 1803

Environmental change is not new. Neither is writing about it. As seen in the above quote, more than 200 years ago the *Scots Magazine* argued for the beauty and the excellence of the Scottish landscape, emphasising the importance of

---

How to cite this book chapter:
Purves, Ross S., Olga Koblet, and Benjamin Adams (2022). "Introduction." In: *Unlocking Environmental Narratives: Towards Understanding Human Environment Interactions through Computational Text Analysis.* Ed. by Ross S. Purves, Olga Koblet, and Benjamin Adams. London: Ubiquity Press, pp. 3–16. DOI: https://doi.org/10.5334/bcs.a. License: CC-BY 4.0

diversity and water bodies, concepts with enduring importance extending far beyond the study of landscape preference into the realm of environmental science and intergovernmental efforts to preserve biodiversity through policy and scientifically driven interventions (Díaz et al., 2015).

The author also bemoans the quality of the land and the lack of forest, and lays the blame squarely at the door of the owners of Scotland's great estates and their improvements. These ideas are familiar in current debates about the environment, as argued for in sustainable development goals that aim to improve food security (and thus agricultural productivity) and the sustainable management of forests. That a 200-year-old quote fits this argument so well is, of course, not coincidence – the way in which humans value and appreciate landscape has a history – in this case a profoundly Western one, which influences to this day relationships with the environment (Fletcher et al., 2021).

Understanding these relationships is central to many of the most pressing environmental challenges we face today. Beauty is not simply in the eye of the beholder, but neither are its manifestations universal, and this variation has profound policy implications if we wish to understand, preserve and enhance landscape quality. Reforestation and rewilding programs stand or fall not simply as the result of legislation promoting these practices and protecting environments, but also through an understanding of the relationship between those living in a region and its history (Root-Bernstein, Gooden, and Boyes, 2018). Environmental change occurs not in isolation, but as a complex interplay between humans and environmental consequences, manifested in processes such as climate change, agricultural improvement or urbanisation.

One rich source of material on the relationships between society, culture and the environment is natural language as recorded in written texts. How we perceive our environment, how we understand the relationships between the behavior of people and the landscape and how our conceptions and discourse about the environment in turn cause change is conveyed through natural language, or what we shall call in this book, *environmental narratives*.

We define environmental narratives broadly for our purpose, as written texts focusing on the environment, often describing events and experiences that took place at a particular location or in a particular type of location. These narratives can take many forms, ranging from non-fiction, through fictional accounts to poetry. As narratives, we explicitly embrace the idea that these texts reflect – often implicitly – a particular position. The extract from the *Scots Magazine* vividly reflects these ideas, describing as it does a particular place at a particular time with a very clear opinion. Combining this account, together with other contemporary accounts of life in Scotland, would reveal a complex synthesis of information containing competing, or even contradictory accounts of concerns about people and the environment in the same locations. Taken together such texts are a powerful way of creating bottom-up pictures of people, places and environments from multiple perspectives.

Environmental narratives have encoded potential answers to questions about the interactions of people and their surroundings for millennia and are as old as literature itself. The concerns of human societies that play out in the written record are set out in accounts of historical events and the settings in which they take place. Thus, when the Greek historian Thucydides wrote about the Peloponnesian War, the arrangement of objects in the landscape played a critical role in his explanation (Thucydides, 1998). The choice of objects made in these descriptions tells us something about what Thucydides, and by extension his audience, considered salient parts of the environment.

Traditionally, the reading and interpretation of such texts has been primarily the domain of interpretative, qualitative methods stemming from the social sciences. These approaches are incredibly powerful. For example, environmental historian William Cronon showed how narratives shape our understanding of the Dust Bowl, an event that had significant influences on policy in the United States and beyond for decades (Cronon, 1992). Interest in reading and interpreting environmental narratives is diverse, spanning areas of inquiry including landscape characterisation, environmental history, tourism studies and investigations of the links between climate and society.

We found the *Scots Magazine* text that opens this introduction using a search engine, and the use of such techniques lies at the core of this book. In the last 20 years academic research and society have been transformed by digital access to, and processing of, enormous volumes of data. Environmental science has been revolutionised by opportunities arising from terabytes of remotely sensed data capturing physically measurable properties. Such research is perhaps best exemplified by works measuring global forest cover and, taken to extremes, aiming to count the number of trees found on Earth (Crowther et al., 2015). But these remote sensing data can tell us nothing about what people call these trees, or how the ways in which trees are valued have and do change in time and space. Answering such questions about humans and their relationship with the environment is, we believe, a challenge to which environmental narratives are ideally suited and the subject of this book.

This book takes advantage of two key developments. Firstly, vastly more texts are openly available and searchable online (Michel et al., 2011). This access to text provides us with an opportunity – it is, at least if we are in need of an apposite quote, much easier to browse huge volumes of text and identify those fitting our needs. But with this opportunity comes a challenge – the identification and interpretation of the quote as being relevant to those needs remains an interpretative task. Given the vast volumes of text available, how can we move beyond identifying single quotes to synthesising environmental narratives so that we can really understand the competing stories about a place?

The second development concerns the development of methods to computationally process text. Many of us use search engines as digital assistants, helping us to find and filter information with regard to not just research, but also making choices in our everyday lives. We have a subjective idea about what sorts of information such systems seem well suited to dealing with (locations of nearby furniture stores) and where they more often fall down (differentiating between song titles preferred by a 7-year-old using their parent's Spotify account and those favoured by the parent). We use natural language, most often in the form of typed words and phrases, but increasingly through the spoken word to interact with such systems, and it just seems to work.

Much of this magic concerns the development of methods in natural language processing. These techniques are concerned with developing tools which allow us to process and extract information from text. It consists of a very large variety of challenges, ranging from the seemingly simple (e.g., dividing a sentence into its constituent words) through the ubiquitous (automatically translating text from one language to another) to the incredibly challenging (doing such translations in such a way that a native speaker does not notice). These tools have moved from being the preserve of a small number of specialists to building blocks for research across a wide range of scientific areas, and their availability and relative ease of use is central to the digital analysis of text.

The astute reader will have noticed that a great deal of the proceeding argument seems somewhat familiar. It reflects many developments in what have been termed the Digital Humanities, fueled by methods and their application across a wide array of traditionally qualitative and interpretative research fields as digital data have become available (Berry, 2012). However, the focus on methods and computation in the Digital Humanities also marks the point of departure of this book. Our starting point is our observation that written text can tell us much about interactions between humans and the environment. Taking as read that sources and methods are available, we argue that the first task in developing approaches to the computational analysis of environmental narratives is to explore how different disciplines can read and interpret such texts. Only by understanding this, can we develop research questions with real relevance not only within, but also across disciplines. At the heart of the book lies therefore not primarily methods and data, but rather a simple question: How can we take advantage of progress in computational methods analysing text to better understand human interactions with the environment, through approaches which foreground the analysis of multiple narratives and interdisciplinary interpretations of such texts?

How can we best explore such a question? Interdisciplinarity, in its most effective form, requires a bringing together of scholars with diverse backgrounds, who together can make a contribution which is more than simply the sum of the parts. In our view, at the heart of truly successful research that spans and brings together disciplines lie theoretically and societally relevant questions, answered by bringing to bear appropriate methods. Our approach to discussing and identifying such questions took as a jumping off point a workshop, attended by scholars from the humanities, social, computational and natural sciences. The themes of the workshop were threefold, and laid the foundations for this book.

- The first theme was concerned with (re)identifying recurrent themes related to understanding environments from multiple disciplinary perspectives which give rise to questions with the potential to be explored through text.
- The second theme set out to identify both resources (in the form of textual corpora) and methods which could be used to explore such questions from multiple perspectives.
- The third theme, focused on developing a set of case studies illustrating how a holistic approach, based on joint work by participants, could propose novel ways of understanding and exploring environmental narratives.

In the text which follows we address these themes using three different approaches. Firstly, the core team of authors worked together on material introducing key themes and linking contributions from the other workshop attendees. Secondly, workshop attendees wrote specific short vignettes answering questions we posed and developed in the workshop, setting out their individual perspectives on the analysis and interpretation of environmental narratives. Thirdly, groups of workshop attendees worked together to author individual chapters exploring specific research questions through a broad range of methods.

The book is designed to provide newcomers as well as experienced researchers with a holistic and interdisciplinary foundation to research with environmental narratives. By mixing vignettes and case studies with an introduction to resources and methods, we provide graduate level students with a set of potential questions that can be addressed by exploring environmental narratives, and a toolbox of potential methods. For experienced researchers, we bring together a diverse set of potential research questions and methods focused around using text to better understand the environment, and provide a jumping off point for future research.

To illustrate why we believe environmental narratives are a potentially rich source of material, we now turn to a second example from Scotland, a short newspaper article discussing the building of a new road across Rannoch Moor in the late 1920s. The moor itself is, by international standards small, covering some 130 km^2 and designated as a protected area for both its habitats and scenic beauty. But, equally importantly, Rannoch Moor has been the subject of a rich set of narratives over the years. It forms a backdrop for Alan Breck Stewart's flight in the 19th-century historical novel *Kidnapped* by Robert Louis Stevenson, is traversed by new nature writer Robert Macfarlane in his book *The Wild Places*, and features as a location in the film of the book *Trainspotting* by Irvine Welsh, documenting the lives of those involved in Edinburgh's nineties drug scene. Two important transport links bound the moor, to the east a railway built in the late 19th century, and to the west a road built in the 1930s which broadly follows the path of traditional routes to the south. The article was published on Saturday, 5th of November 1927 in a Scottish newspaper, *The Courier and Advertiser*, without a byline, that is to say we do not know who wrote it.

The Glencoe Road

There are signs that the Ministry of Transport is in for a hot time when Parliament meets over its plans for constructing a great motor highway through the Moor of Rannoch and Glencoe. There is, so far, complete unanimity among those who have written and spoken on the subject that the plan is an outrage on sentiment and good taste. It will ruin the scenery and character of a route which is hardly ever traversed save for the sake of its scenery and character. But the case against the Ministry's plans on the ground of common sense is at least equally strong. Sir John Stirling Maxwell, who probably knows the West Highlands as well as anybody in Scotland, declares in a letter to the *Times* that the present road could have been made good for the traffic that uses it for the sum spent by the Ministry in surveying and designing the new road. That on such a road the Ministry should propose to spend half a million at the time when it has in effect stopped the work on the three-quarters completed West Highland road from Glasgow to Inverness because of shortage of funds is the crowning wonder. A feeling is taking hold that the Ministry is not to be trusted where, as is frequently the case in road-making, good taste is involved. It has irretrievably ruined the beauty of many parts of the Great North road, and its concrete bridges are every-where blots on the landscape and eyesores.

Courier and Advertiser, 5th November 1927

Having read this text, we can answer some simple questions. Something, a road, is to be built somewhere – through the Moor of Rannoch and Glencoe. Someone, Sir John Stirling Maxwell has a strong opinion about the plans of a faceless Ministry, and it appears that the road will be built soon. The reason for the road building is unclear, since no explanation is given as to its purpose. The tone of the article, and by extension its anonymous author, is indignant. What though do scholars from different backgrounds read from this article, and how do they contextualise it? To find out we asked six attendees at our workshop with backgrounds in environmental history, literature, digital humanities, landscape studies and tourism research to interpret the piece briefly in writing. To do so, they were provided with a digital copy of the original newspaper article on the page, giving extra context to which some chose to refer. The individual interpretations are found at the end of this chapter, and in what follows we summarise key themes identified by this interdisciplinary group.

The first two pieces from Sarah Luria and Flurina Wartmann (with backgrounds in English and geography with a focus on landscape research, respectively) focus their interpretations on a more or less direct reading of the article. Flurina points to the article's implicit argument against change, and for preservation of beauty and brings to the fore the unheard voices of those who may travel this route for more utilitarian purposes. Sarah identifies a villain (the government) and the same utilitarian need for rapid travel. She points out two ways in which the author argues – through rationalist and romantic discourses, and suggests ways in which the article's tone reflects (typical) British notions of class. Already, in these first two pieces we see how important different actors, arguments and perspectives are in readings of the piece.

Environmental historian Karen Jones and digital humanities scholar Joanna Taylor both explicitly bring historical context to their interpretations. Karen talks to the Highland Clearances and colonialism more generally, while Joanna places the road firmly in its historical context, linking its predecessor to road building after to the Jacobite rebellion of 1745. Joanna also talks to the changing voices used in the article, pointing out the deliberate use of the passive as a means of (re)claiming objectivity. Like Flurina before her, Karen reminds us of all of the voices not heard or considered in the article and points out the ways in which progress (in terms of modernisation) is seen as something to resist.

Graham Fairclough, a practitioner with a long career at English Heritage and Katrín Lund, a human geographer with a focus on tourism both emphasise identity. In Katrín's case through the landscape's contribution to being Scottish, while Graham explores Scotland's position within an Empire for which the Ministry is a proxy. Both also point to the importance in the article of one man, Sir John Stirling Maxwell, with Graham pointing to his leading role in founding organisations concerned with conservation in Scotland. Once again, the underlying move to modernise and the growth of car use, are given as underlying reasons for the building of the road, and both identify the use of sentimental and rational arguments against change.

What becomes very clear on reading these pieces is the importance of additional contextual knowledge in interpreting the article, and how varied the interpretations are, despite relative unanimity on the key messages. One way of summarising the answers of our experts is to categorise them. In Table 1.1 we do this using the 5Ws & H (what, why, when, where, who and how), a set of questions which have been used to explore and discuss texts since antiquity. This simple structure is a powerful way of summarising the arguments put forward by our team. What is telling are the rich variety of themes identified, not only with respect to the objects described in the article (the landscape, scenery etc.), but also the locations and actors, as well as the influence of historic events on the debate about modernisation. Equally important for our work are the ways in which arguments are advanced – through the use of different forms of language and argument, and by using particular and emotive choices of words.

This example shows vividly the potential of interpreting a single environmental narrative. Those familiar with the computational analysis of text, will see some possible avenues for using such methods to analyse multiple texts describing this location. For example, using named entity recognition it might

Question	Selected answers
What	Landscape Scenery Romantic wild nature Landscape transformation Military roads and droving Bridges
Where	Moor of Rannoch and Glencoe Scotland Scottish North Periphery of the Empire Modern A82
Who	Urban-based intellectual The author (identified as male) The Ministry Modernising national government Sir John Stirling Maxwell Cultivated elite Major William Caulfeild Ordinary folk, women, non-humans Local people Association for the Preservation of Rural Scotland National Trust for Scotland
When	Jacobite Rebellion Highland Clearances Increasing presence of motor car Birth of countryside conservation movement
Why	Modernisation Conservation
How	Passive voice Colloquial language Sentimental Rational Conservative Clichés

Table 1.1: Interpretation of the text Glencoe Road.

be possible to identify the locations and actors explicitly referred to. Of course, much of the interpretation – for example, the lists of missing voices – cannot be extracted computationally. But, and this idea is at the heart of this book, the process of interpreting the texts with experts from diverse backgrounds allows us to identify relevant questions, for which we can then consider potential approaches to analysing text. This example provides us with a jumping off point for the rest of the book and our argument. In the first part of the book, made up of this chapter together with Chapters 2 and 3 we focus on exploring firstly, what environmental narratives can tell us (our first theme) (Chapters 1 and 2). In Chapter 3, we introduce both resources (collections of text and associated data) and simple methods for text analysis that might be appropriate to analyse some of the questions we posed above. Chapter 3 is designed to give those new to text analysis an insight into possible approaches, and could be skipped by those familiar with such methods.

In the second part of the book, we focus on case studies developed by the attendees of our workshop. Each case study team (with one exception) was interdisciplinary, and the resulting examples illustrate how a diverse range of questions, collections and methods can help us to explore environmental narratives. The book concludes with a list of key lessons and sets out an interdisciplinary research agenda for future work.

Interpretations of the Glencoe Road text

Walking the Glencoe Road as an environmental historian the sign-posts are many. Here we detect a long story of landscape trans-formation, from the Highland clearances to the construction of a colonial landscape of romantic wild nature. Movement and mapping are strong themes in this place – from military roads and droving to modern auto tourism. Axioms of progress courtesy of industrial technology are embedded in the text – the motorway as a beacon of modernism – along with a critical inflection that sees the 'machine in the garden' as bureaucratic invader and aesthetic polluter. What is missing, of course, from the Courier's report are the other voices – the ordinary folk, the women, the non-humans – important, hidden, historical actors in this story.

Karen Jones
School of History, University of Kent, UK

The Glencoe Road – the modern A82 – is composed of a series of historic routes which have been in use since at least the 18th century. Like several parts of this road, the section which runs by Rannoch Moor formed part of the way constructed under the leadership of Major William Caulfeild, who was Inspector of

Roads for Scotland from 1732 until his death in 1767. He was thus in charge of Scottish transport links when the second Jacobean War was underway in 1745. The Dundee Courier and Advertiser registers something of this historical antagonism against English administration in Scotland. The author's opinion that the Ministry is 'in for a hot time' indicates a muted delight at this outcome; the colloquial language with which the article opens is at odds with the spitting fury later on, when the author bemoans a plan which is little more than 'an outrage to sentiment and good taste', and which, furthermore, is barely defensible on practical grounds. The passive voice which opens the final paragraph ('a feeling is taking hold') attempts to reclaim a sense of objectivity – but the paper's stance is clear. Its position is reinforced by the placement of this article on the broadsheet: this is a central story, positioned in the middle of a page dominated by international news of places from Abyssinia, to Egypt, to Hollywood. As both the article itself and its location on the page indicate, though, this issue is no less significant for either newspaper, author or reader because it is a local concern. The 'scenery and character' of the landscape through which the road will pass stands in for complex concerns around Scottish heritage and identity. Both, the article fears, will be significantly harms if the Ministry gets its way.

Joanna E. Taylor
Art History and Cultural Practices, University of Manchester, UK

The Ministry of Transport is going to have some trouble initiating the construction of a new road through Glencoe. This is not because of some sentimental voices that have frequently been expressed in writing, which talk about the scenery that will be ruined but rather because a man of knowledge and reason, Sir John Stirling Maxwell, has put forward rational arguments against it.

Whilst the Ministry seems to want to head into modernity with the road construction the voices against, both sentimental and rational, are conservative and seek to protect the landscape in as original shape as possible for the sake of its beauty, which in my interpretation seems to be a landscape that represent what it means to be Scottish, at least for some.

Katrín Anna Lund
Institute of Life and Environmental Sciences,
University of Iceland, Iceland

The text provides a viewpoint of a (presumably) urban-based intellectual and his peers that the Scottish North, and particularly Rannoch Moor, should be preserved in its current state for the appreciation of people such as the author. Landscape is seen as a scenery to be driven through and taken in with people of 'good taste'. What is interesting is that the historic significance of the landscape is not mentioned, which could also have been presented as a reason against road development. Furthermore, concrete bridges are portrayed as 'blots on the landscape and eyesores', rather than as a necessity to connect settlements. The views of local people who may use the road and bridges not for sightseeing but more pragmatic needs such as transport of people and goods are not mentioned at all.

Flurina Wartmann
University of Aberdeen, UK

The villain in this editorial is a modernising national government, whose 'great motor highways' destroy landscapes by treating rural space as something to traverse as quickly as possible, presumably to get from one important (urban) center to another. According to the author, the proposed road will blanket the current smaller route and 'ruin' a landscape whose use value is precisely 'its scenery and character'. To persuade the reader of his superior view, the author invokes a rationalist discourse of 'common sense', and a romantic discourse that prioritises 'sentiment', 'scenery', and 'beauty'. This is further augmented through the author's intimidating quasi-upper class tone of outrage at the assault of 'good taste' by the 'blots' and 'eyesores' introduced by modernity onto a landscape best suited for romantic leisurely perambulation and aesthetic appreciation. These rhetorical coercions aside, the author correctly points to the fundamental connection between routes and landscapes: the bigger the road the more impacted the surrounding landscape. Eight-lane highways create backbones for sprawl. Thus, road design and building should continue to be 'hotly' debated.

Sarah Luria
Environmental Studies/English Dept., College of the Holy Cross, Worcester, MA

This article from a Scottish newspaper only has two arguments explicitly against the road, neither strong: damage to good taste and waste of public money. It is not fully anti-road but it does envisage the 'ruination' of scenery and character. The case against the roads construction however barely goes beyond a few charged, and even in 1927 clichéd, terms, e.g., 'blots on the landscape', 'eyesores', and the road is to be built 'through' Rannoch Moor and Glencoe rather than (less dramatically) 'across and along'.

In contrast, the article's principal criticism of the planned road is that it is an outrage on 'sentiment and good taste' (which are essentially the preserve of a cultivated elite). The 'complete unanimity' which (of course) those terms frame is concretely symbolised by the authority of the 10th Baronet of Pollok, Sir John Stirling Maxwell (1866–1956), the man lauded by the Courier's journalist as the person who knows as much as anyone. A sense of value is only evident in money terms, through the 'common sense' opinion that public spending can be cut by a cheaper improvement of the existing Old Glencoe Road to make it just sufficient 'for the traffic that uses it' (there is little understanding yet that roads create demand).

This is a political narrative, close to the heart of the article, from the first sentence's finger-pointing to London to the closing paragraph's delicate sarcasm of 'A feeling is taking hold that the Ministry is not to be trusted….' and its negative comparison with the Great North Road. Spatially, we are in the periphery of a distrusted Empire, not in the wilderness of Rannoch, and politically we are in a conservative narrative (ironically an editorially critical immediately adjacent column in the newspaper is headed 'What is Socialism?').

Temporally, 1927 places us at two slow turning points. First, is the increasing presence of the motor car (which is visible at the top of the same page of the Courier in the contents box advertising the 'Scottish Motor Show') with growing worries about how to exploit, regulate, facilitate and control it. Second, is the birth of the countryside conservation movement: in 1926 the Association for the Preservation of Rural Scotland (APRS) was founded and in 1931 the National Trust for Scotland (NTS), both co-founded by Sir John Stirling Maxwell.

Graham Fairclough
School of History, Classics and Archaeology,
Newcastle University, UK

References

Berry, David M (2012). "Introduction: Understanding the digital humanities". In: *Understanding digital humanities*. Springer, pp. 1–20. DOI: 10.1057/9780230371934_1.

Cronon, William (1992). "A place for stories: Nature, history, and narrative". In: *The Journal of American History* 78.4, pp. 1347–1376. DOI: 10.2307/2079346.

Crowther, Thomas W, Henry B Glick, Kristofer R Covey, Charlie Bettigole, Daniel S Maynard, Stephen M Thomas, Jeffrey R Smith, Gregor Hintler, Marlyse C Duguid, Giuseppe Amatulli, et al. (2015). "Mapping tree density at a global scale". In: *Nature* 525.7568, pp. 201–205. DOI: 10.1038/nature14967.

Díaz, Sandra, Sebsebe Demissew, Julia Carabias, Carlos Joly, Mark Lonsdale, Neville Ash, Anne Larigauderie, Jay Ram Adhikari, Salvatore Arico, András Báldi, et al. (2015). "The IPBES conceptual framework–connecting nature and people". In: *Current Opinion in Environmental Sustainability* 14, pp. 1–16. DOI: 10.1016/j.cosust.2014.11.002.

Fletcher, Michael-Shawn, Rebecca Hamilton, Wolfram Dressler, and Lisa Palmer (2021). "Indigenous knowledge and the shackles of wilderness". In: *Proceedings of the National Academy of Sciences* 118.40. DOI: 10.1073/pnas.2022218118.

Michel, Jean-Baptiste, Yuan Kui Shen, Aviva Presser Aiden, Adrian Veres, Matthew K Gray, Google Books Team, Joseph P Pickett, Dale Hoiberg, Dan Clancy, Peter Norvig, et al. (2011). "Quantitative analysis of culture using millions of digitized books". In: *Science* 331.6014, pp. 176–182. DOI: 10.1126/science.1199644.

Root-Bernstein, Meredith, Jennifer Gooden, and Alison Boyes (2018). "Rewilding in practice: Projects and policy". In: *Geoforum* 97, pp. 292–304. DOI: 10.1016/j.geoforum.2018.09.017.

Thucydides (1998). *History of the Peloponessian War*. Trans. by Steven Lattimore. Indianapolis: Hackett Publishing Company, Inc.

CHAPTER 2

Interdisciplinary Perspectives on Environmental Narratives

Ross S. Purves
Department of Geography; URPP Language and Space, University of Zurich,
Switzerland

Olga Koblet
Department of Geography, University of Zurich, Switzerland

Benjamin Adams
Department of Computer Science and Software Engineering, University of Canterbury,
Christchurch, New Zealand

Katrín Anna Lund
Institute of Life and Environmental Sciences, University of Iceland, Iceland

Karen Jones
School of History, University of Kent, UK

Sarah Luria
Department of English, College of the Holy Cross, Worcester, Massachusetts, USA

Flurina Wartmann
Department of Geography & Environment, University of Aberdeen, UK

Graham Fairclough
McCord Centre for Landscape, Newcastle University, UK

Gabriel Viehhauser
Department of Digital Humanities, University of Stuttgart, Germany

How to cite this book chapter:
Purves, Ross S., Olga Koblet, Benjamin Adams, Katrín Anna Lund, Karen Jones, Sarah
Luria, Flurina Wartmann, Graham Fairclough, and Gabriel Viehhauser (2022).
"Interdisciplinary Perspectives on Environmental Narratives." In: *Unlocking
Environmental Narratives: Towards Understanding Human Environment
Interactions through Computational Text Analysis.* Ed. by Ross S. Purves,
Olga Koblet, and Benjamin Adams. London: Ubiquity Press, pp. 17–42.
DOI: https://doi.org/10.5334/bcs.b. License: CC-BY 4.0

This book is about unlocking environmental narratives by using computational methods to extract and analyse information from text. But what do we mean by *environmental narratives*, and just as importantly, what are different ways of thinking about them? The following six vignettes explore these ideas from widely varying academic perspectives. They vividly illustrate not only the breadth of forms of environmental narrative, but just as importantly the widely varying ways in which these narratives can be explored.

Our first vignette, by human geographer Katrín Anna Lund, tells the story of a particular landscape, that of Strandir in northwest Iceland from a phenomenological perspective. In doing so, Katrín tells the story of Strandir and its landscapes from her viewpoint as an Icelandic woman and human geographer, travelling through the region, with a knowledge of its history and the environment. She shows how landscape is on the one hand in the eye of the beholder, but on the other a shared product of history and environment. Katrín's vignette is important because it shows through a specific example the value of narrative in understanding landscape. It is an individual piece, typical of work in human geography, which through its use of the first person makes its positionality clear. Worth considering, is how this piece of writing might influence what we would see if we visited, or worked on, materials related to this landscape in the future.

Strandir Stories: A phenomenological approach to narrative

I use the region of Strandir in northwest Iceland to reflect on what Rose and Wylie (2006) mean when they talk about 'landscape as tension'. My approach is phenomenological, exploring landscape from the standpoint of a lived and perceiving body (Merleau-Ponty, 2002) experiencing its surroundings as it moves through them. Furthermore, as Bender (2002, p. S106) put it, landscapes 'refuse to be disciplined' – they are never passive and still, but they play upon us, move and affect us (Lund and Benediktsson, 2010).

When travelling through the Strandir region one cannot but feel the strong connections to the environment even if travelling by car, which is the usual mode of transport in the area. Following the narrow gravel road allows the traveller to physically feel the rugged and barren surroundings as it winds along the coastline and over mountain sides connecting fjords and places (Figure 2.1a). The population is low – the village Hólmavík has about 450 inhabitants, whilst the remaining 300 or so inhabitants are scattered over the region mostly living on farms or in small hamlets. There is a strong sense of remoteness, rooted in a timeless entanglement of present and past that provide a

(a) The road.

(b) Weather.

(c) Driftwood.

(d) Eider duck, nesting.

(e) Puffin.

(f) Museum.

(g) Petrified trolls.

Figure 2.1: Environmental features of the Strandir region. 2.1b by Claus Sterneck, all other pictures by Katrín Lund.

sense for a still calmness, nevertheless in a landscape that is ever moving and playful as it stirs up narratives which take the traveller into different spatio/temporal dimensions (Lund and Jóhannesson, 2014). As the road goes along the narrow coastline, stones and the occasional rock on the road remind the traveller that the mountains towering above are ever moving in their battle with the forces of weather, wind, rain and snow and simultaneously that life in the region has always been a struggle with unpredictable nature (Figure 2.1b). Along the coast line lie unruly piles of driftwood, carried by ocean currents from distant Siberia. Driftwood (Figure 2.1c) used to be the region's most valuable treasure in a land otherwise barren of wood and stirs up thoughts about how ever moving natural forces simultaneously give and take. The same can be said about the eider duck (Figure 2.1d) swimming by the shore amongst other migrating seabirds (Figure 2.1e). Its down has always been a highly valued subsistence for farmers in the area (Lund and Jóhannesson, 2014). This brings forth visions into the past, and not the least to the 17th century. Men in authority, usually in the name of the Lutheran establishment, desired these valuables. In turn, this competition for resources was one reason behind witchcraft accusations and hunts, a piece of history that the area is still notorious for, reflected in the founding in 2000 of the Museum of Icelandic Witchcraft and Sorcery in Hómavík (Figure 2.1f). The opening of the museum stirred up otherwise almost forgotten narratives, so that the landscape can be sensed as magical (Lund, 2015; Lund and Jóhannesson, 2016). These magical or unearthly experiences are also reflected in stories of mystical beings such as trolls, elves and sea monsters. Rock formations at the shoreline are often said to be the remains of trolls that did not take care and turned to stone when they were caught by the rays of the sun (Figure 2.1g).

Thus the landscape experienced is moulded by the comings and goings of people, earthly materials, animals and supernatural beings travelling to, from and within the region, stirring up the narratives that enmesh the traveller. In so doing, we bring own narratives and stir up others. This landscape is more-than-human, it is a landscape that is vital, constantly moving, and tensioned; a landscape where nature and culture refuse to be separated and the traveller is part of the story.

The hamlet of Djúpavík reveals how places are gatherings of narratives stemming from more-than-human mobilities (Figure 2.2). The ruins of an old herring factory dominate the surroundings (Figure 2.3a). The factory was built in 1934, paid for by venture

Figure 2.2: Djúpavík. Picture by Claus Sterneck.

capital brought in by entrepreneurs from the Capital, with the aim
of becoming wealthy through the abundance of herring that had
arrived in the surrounding seas. Djúpavík became a thriving place
where people from all over Iceland gathered for work every summer during the herring season. However the adventure only lasted
about 20 years before the herring moved on (Lund and Jóhannesson, 2014).

In 1975, a young couple from the Capital arrived in the now
deserted place. They decided to settle and renovated an old building, still called The Female Quarter, as it housed the women working in the factory, showing how placenames can fix the past in
the landscape. They also established a small hotel, still in oper-
ation today, which attracts visiting travellers. After a slow start,
business has boomed, especially during the tourist seasons, with
visitors now coming all year round. More-than-human narratives emerge through the ruins of the factory and, more over,
how they are continuously in the making. They are maintained
by the owners to avoid danger to curious visitors, and they have
installed an exhibition about the history of the factory inside it
(Figure 2.3b).

(a) The factory.

(b) The exhibition.

(c) From inside of the factory.

(d) Factory's windows.

Figure 2.3: Environmental features of Djúpavík. Pictures by Katrín Lund.

An art gallery, a storage room for diverse things that belonged to the place's past including old vehicles restored by the owner in his spare time (Figure 2.3c) add to its story. Non-human material processes also contribute to the narrative, as the forces of water and wind are continuously working their way through the concrete, appearing in the form of mould and moss shaping living spaces for insects, birds and rodents (Figure 2.3d). The factory ruins demonstrate how human and non-human forces become entangled and weaves together diverse spatio/temporal narratives.

Katrín Anna Lund
Institute of Life and Environmental Sciences, University of Iceland,
Iceland

Our first vignette used phenomenology to write and present a narrative of a specific landscape. It vividly illustrated how words and pictures can be combined to help us imagine a landscape, and categorise it. Katrín emphasised the remoteness of this landscape, but nonetheless concentrated on the interplay between people and natural processes in our understandings. This is no empty landscape, no wilderness, but a place where wind and weather erode stones and cliffs that are suffused with stories from the past. But where do notions of wilderness, and nature itself come from, and why are they more complicated than we sometimes think?

Our second vignette, from environmental historian Karen Jones, delves deeper into the origins of the words and ideas we use to characterise landscapes. In doing so Karen takes us on a journey through nature, into the wilderness and finally for a walk in the park. She shows how our ways of understanding landscape cannot be separated from their history, and the history of the words we use to talk and write about landscapes.

Entangled words and materials: Environmental history and an etymo-cology of meaning

Nature, Raymond Williams tells us, is perhaps the most complex word in language (Williams, 1983, p. 219). There are, as it turns out, many contenders in a fecund modern environmental vernacular. From *Anthropocene to Wilderness*, the words we use to categorise physical spaces, processes and interactions are layered with imaginative significance and, some might say, hamstrung by a messy and (perhaps inevitably) problematic provenance. This makes the matter of narrative particularly significant. Indeed, by digging a bit deeper into the world of etymo-cology (a term I use here to describe a forensic enquiry into words and matter, imagined and material traces) we discover a landscape that is contested, controversial and eminently more intriguing for it.

Environmental history provides an important guide in navigating these entanglements of materiality and cultural meaning. As Stephen Dovers notes, 'an environmental issue without a past is altogether as mysterious as a person without a past'. (Dovers, 1994, p. 4).

At its core, the discipline announces that human experience cannot be read in isolation from the physical world. Exponents argue that history has been too anthropocentric, of a need to put nature back into history (or the other way round). As practice, it has tended towards three kinds of enquiries: 1) How nature has changed over time; 2) human environmental impacts; and 3) cultural representations, values and ethics (Worster, 1988, p. 293). The field gained momentum as part of the revisionist drive to make history more inclusive, with early exponents focusing on (deleterious) anthropogenic impacts on the biosphere and tracking the roots of environmental consciousness. This genealogy is important, as despite the fact that the discipline has matured into a sophisticated canon of eco-cultural enquiry, an activist element still remains important. Where does environmental history end environmental-*ist* history begin? Such responsibilities are particularly prescient in our contemporary world of plastic overload and planetary crisis. As Ruth Morgan notes, climate change has firm implications for the 'ways in which we undertake writing history' (Morgan, 2013, p. 350).

Going back to the nomenclature of nature, one of the critical terms that has been forensically examined has been 'wilderness'. In *Uncommon Ground* (1995), environmental historian William Cronon invites a rethinking of the idea to reveal a landscape of social construction, an ethnically vantaged fantasy, and a consumer product. Read in this vein, 'the wild' becomes a place of escape, a paradise untouched by industrialism, and something for sale in the mall. Advising humans to find a home *in* nature, a common ground in which to live responsibly, Cronon's argument ably highlights academic writing as activism. As he asserts, 'the special task of environmental history is to assert that stories about the past are better, all other things being equal, if they increase our attention to nature and the place of people within it'. (Cronon, 1992; Cronon, 1995).

For the rest of this vignette, I want to showcase the potential of environmental history to unpack environmental narratives via a short walk around 'the park'. Here, too, we discover a rich palette of shifting meanings. In its original definition, 'Park' described an enclosed piece of ground for the beasts of the chase. Boasting a distinguished lineage, so-named places encompassed the hunting preserves of ancient Assyria, medieval deer reserves of European royalty, and Versailles, Louis XVI's geometric hydraulic masterpiece that mapped the power of the Sun King over nature and nation. While appearing to show the emergence of a more 'natural' park variant, the pastoral lines of the English landscape park, popular in the 18th century, was just as much a designed entity. It also confirmed the importance of the park as a prime site of narrative. Stourhead in Wiltshire, the brainchild of banker Henry Hoare II, clearly depicted syncretic lines of landscape and storytelling: its circular walk around a lake, complete with shaded walk and river grotto, telling the legend of Aeneas' journey into the underworld.

The 19th-century cityscape communicated a narrative of civic progress through industrialism. Depicted in a canvas by William Wyld, 'Manchester from Kersal Moor, 1857' inferred that city and nature could sit comfortably, with romantic hills and industrious smoke stacks presented as symphonic. Over time, however, the social, economic and environmental consequences of industrialism raised concerns about community health. It was in this context that the urban park idea emerged: a rustic ideal-type transplanted to the city to make it liveable. Attributed to William Pitt and first cited in parliamentary debates about urban development on the edges of London's Hyde Park, the park as 'lungs for the city' positioned it as a kind of environmental and social prescription, a green enclave that circulated the air and allowed citizens to walk and breathe healthily. Also worth noting was the importance of social engineering in these 'parks for the people'. Ordered by racial, gender and class-based codes, these were spaces of socialisation and orderly recreation. As leading American landscape architect Frederick Law Olmsted noted: 'if thousands of people are to seek their recreation…unrestrainedly, each according to his own special tastes….[the park]…is likely to lose whatever of natural charm you first saw in it' (Olmsted, 1881, p. 26).

Most famous in the dedication of Yellowstone (1872), the *national* park communicated a story of scenic grandeur and stood as an important marker of the celebration of 'the wild'. Suggesting firm connections between the veneration of nature, cultural nationalism, and the emergence of a conservationist ethic, Yellowstone was preserved in 'natural conditions … for the benefit and enjoyment of the people'. The notion of what a national park was *for*, however, changed radically over time. Here the collisions of etymo-cology were clearly seen as managers and visitors grappled with such issues as indigenous access to park lands; the role of fire regimes in a protected landscape; and the place of 'good' and 'bad' animals, a tension most famously seen in the eradication, and later reintroduction, of wolves.

Parks represent complex spaces. What defines them? What is *their* narrative? How do push-button geyser simulations in Disneyworld complicate the story of Yellowstone's 'pristine' Rocky Mountain nature? What of somewhere like Pripyat amusement park, Ukraine, abandoned to its former use after the explosion at nearby Chernobyl, and now an unintended nature reserve roamed by wild boar, brown bear and wolves (largely due to the absence of people)? As a body – or corpus – of material, 'the park' presents an avalanche of contested stories both *in* and *of* the land. Trying to make sense of this complex landscape of matter and meaning is usefully aided by environmental history methods that track the lines of physical change and can 'read' a range of different landscape texts to show processes of imaginative re-mapping at work. This kind of etymo-cological excavation, I'd argue, is essential for understanding both time and place in environmental narratives.

Karen Jones
School of History, University of Kent, UK

Karen's vignette shows how delving into environmental histories through text can reveal influences, propagated over time that shape our understanding of landscape. Often, these concepts, conveyed by words like 'naturalness' or 'wildness', are used without the writer necessarily being aware of their complex histories. But words can also be used deliberately, to change the way we think about a place.

Our third vignette, from literary scholar Sarah Luria explores a very different setting, the metamorphosis of the industrial city of Worcester in Massachusetts to a new, gentrified and green town. Building on the vignettes from Katrín Lund and Karen Jones, Sarah takes a narrative approach, analysing the stories told by two observers and protagonists of change – a developer, working to revitalise the area and a poet, documenting vanishing ways of life. Sarah's approach complements those of Katrín and Karen, using written and oral sources to explore how change in Worcester is described, and how this narrative is contested by contrasting stories seeking to privilege the past and the future according to the needs of the protagonists.

The power of Blarney: Reinventing an old industrial city as a green mixed class community

A city is not only its buildings, streets and the people who live there, but the stories they tell. The old industrial city of Worcester, Massachusetts, demonstrates this well. Once known as the 'Dirty Woo', today Worcester has been dubbed the 'New "It" Town', an 'authentic' historic city that is being 'revitalised' for modern greener lives, its boarded up factories repurposed for trendy lofts, cafes and bars (Schacter, Aaron, 2018). Young professionals are moving in, attracted by a walking city lifestyle and improved commuter service from handsome Union Station into Boston. A familiar story of gentrification seems to be unfolding, especially in the city's Canal District, where current low income residents may soon be priced out. Two powerful local storytellers, one an upscale developer, the other a working-class poet, capture this conflict and model ways to rewrite it.

Beginning around 1980, developer Alan Fletcher and a group of men met monthly at the 3G Sports Bar on Millbury Street in the Canal District to swap visions of how Worcester could be brought back to life. They became obsessed with the Blackstone Canal that once flowed right by the neighbourhood's busy factories to the Atlantic Ocean 45 miles away, but now lies buried underground in a sewer pipe. Fletcher and his friends dreamed of recreating the canal with boats lined by shops and parks. They did more than just talk. They formed the Canal District Alliance and broadcast views of what the neighborhood had been and could be: they sponsored two huge murals on local buildings with romantic depictions of the old canal and immigrant neighbourhood, a walking tour of historic sites, and a feasibility study for the canal park with an alluring video of how it would look when done.

Fletcher's success suggests key strategies for how to revive a struggling neighborhood: have a clear single focus, a history-based story told through multiple genres, and most importantly, think big. Recreating the canal turned out to be pure fantasy, but, as Fletcher tells it, that did not matter. The alluring vision was enough to attract investment:

I could probably name 10 to 20 people who came to those [bar] meetings, got infected by our bullshit, entranced by the pictures we were weaving in the air, and thought, 'Yea, I've always wanted to open my own bar,' and they bought a place in the Canal District... That's when the rebirth of the Canal District, which started as a bar scene, happened.... I attribute that, very much, to all the yakking we did.

As one local reporter put it, the current success of District is 'fueled' not by a canal of water but a 'river of booze' (Quinn, Tom, 2016).

Fletcher 'take[s] comfort' in the 'fact' that the escalating development isn't 'displacing anybody here in the main part of the district, because nobody lived here anyway.' This, however, is another fantasy story, commonly used to justify gentrification. In fact, a working-class immigrant community, now largely Latino, still lives in the Canal District, and may soon be displaced (Smith, 1996; Hibbett, Maia, 2019).

Former Worcester resident Mary Fell offers a counter-narrative to Fletcher as well as meaningful connections to the local history Fletcher claims to want to preserve. Fell, a working-class Irish-American poet, documented the neighborhood when it touched bottom during its decline. Her 1984 poem 'Prophecy'[1] tells a story of neighborhood identity and ownership that bears repeating today. It begins:

The old neighborhood remains. Some call it Green Island, remembering the canal that cut through it, now underground. Built by Irish laborers, the canal gave Water Street its name. Jews still run their shops there, though they've moved their families to the other side of town. No one goes down in the basements anymore. Rats the size of dogs, they say. Kids in the Catholic school learn Polish prayers. And on Millbury and Harding Streets, everyone talks big stories in the same old bars (Fell, 1984).

Fell's first stanza counters Fletcher's romantic murals of the canal district's past. The short first sentence is richly ambiguous: its flat tone suggests a depressing neighborhood of people left behind but also sounds resilient – the neighborhood has survived. Children live here and still learn Polish, and parents maintain their ethnic pride. The ending even links Fletcher's pals' 'bullshitting' to the 'big stories' – Fell's Irish would call it blarney—that everyone still talks in 'the same old bars'.

In fact, Fell's poem shows how today's neighbourhood has long been supported by alcohol. Worcester historian Roy Rosenzweig notes that alcohol offered relief from the pressures of factory work. Workers slowly won shorter work days and better wages, and enjoyed their leisure in local pubs. These included the 'shebeen', or woman-owned kitchen distilleries, a tradition brought from Ireland. Fell includes a vivid picture of one in her poem. The speaker's grandmother Aggie made:

whiskey in her own still. Though she and her husband were American born, Patsy spoke all his life with a slight brogue. Winter nights, the cop on the beat would come in from the cold to warm himself with Aggie's brew. Putting his little glass under the still, where whiskey squeezed into it drop by drop, he'd run out to the callbox to tell the station all was well. When he got back the glass would be full.

The shebeens gave women, and widows especially, a way to support themselves. But as the city regulated alcohol, they were shut

---

[1]   Stanzas from the poem "The Prophecy" by Mary Fell have been used with permission of the author. All rights reserved for all elements of the poem.

down in favour of male-owned public saloons (Rosenzweig, 1983, p. 43-44).

Fell's vignette should be quoted in today's Canal District bars to honor the Irish establishments that preceded them. As Fell's "Prophecy" demonstrates, maintaining a neighborhood's connections to its longer history is essential to preserving its soul. Aggie and Patsy maintain their Irish roots with the shebeen, Patsy's Irish brogue, and their stories. Her father however has lost that connection:

Though my father is a storyteller, he has little else to say. When he was ten his mother died. He can't remember his grandparents or their names. A few photographs survive, some unidentified.

The poem ends again ambiguously: the depressing note of the father's silence is contradicted by the poem itself, which preserves the neighbourhood's Irish past by inheriting and telling its story.

Fell's poem is a prophecy: the old neighbourhood must still remain. Stories require a past, present and glimpse of the future or else there is no story, no meaning. That past is embodied not only by its buildings and historic plaques, but by residents who remember or have some connection to its past.

Communities change, but gentrification is increasingly acknowledged as inhumane. Fell and Fletcher demonstrate the power of an alluring story well told (the kitchen pub, the buried canal) to change a neighborhood in memory and in fact. Right now we crave a better story than gentrification, which eviscerates the authentic neighborhoods it seeks. The Canal District may have the roots to realise such a fantastic story, one that includes new businesses, a mix of social classes and ethnicities, and a respect for the past and community nurtured through its many pubs.

Sarah Luria
Environmental Studies/English Dept., College of the Holy Cross, Worcester, MA

Our first three vignettes explore environmental narratives as sources for academic inquiry. But how is narrative, or writing about the environment linked to policy? Or, turning the question around, how are the multitude of voices and stories which shape our understanding of environments (and more specifically landscape) currently represented in landscape monitoring? The next two vignettes present two complementary, but very different views.

The first, from Flurina Wartmann introduces the notion of indicator-based methods, often used to monitor landscape and policy success. The second, from Graham Fairclough, presents two related approaches developed in the UK, Historical Landscape Characterisation and Landscape Character Assessment (LCA). These methods are specific examples of the more general notion of indicators presented by Flurina, and both of them aim to tell the story of specific landscapes, through very different data sources, methods and ways of communicating their results. These vignettes introduce new concepts, such as landscape monitoring, ecosystem services and landscape management and do so from a different position – that of directly making inputs into landscape policy and management. In doing so, they emphasise the breadth of knowledge and ideas that are important in work on environmental narratives, and once again emphasise the need for interdisciplinary perspectives to bring together the necessary knowledge to work effectively on these topics.

Landscape and policy

Modern policy emphasises the importance of everyday landscapes where we live and work and, as these become increasingly urban, the role of urban green spaces and rural landscapes as places for recreation and relaxation. Landscape is thus key to developing sustainability policies since it integrates environmental, social and economic aspects of modern society. Many landscapes have been transformed through generations of human settlement and use into cultural landscapes through human actions, ranging from rural forest landscapes used for timber and protection, through extensively and intensively farmed landscapes to bustling cityscapes. Irrespective of their nature, people form strong bonds and connect with landscapes, imbuing places within them with cultural meaning. Landscapes and the cultural values associated with them thus form an essential part of people's relation to place and identity.

Many different processes impact on landscapes, and they are thus constantly in flux. Changes may be rapid, driven by processes such as a mudslides or development of new neighbourhoods. Equally, slower, and less immediately perceptible changes occur with the slow creep of tree lines upwards as a consequence of climate change, or the gradual process of agricultural intensification and associated loss of small-scale landscape structures including hedgerows. From a policy perspective, it is important to document and monitor such changes, since they influence physical landscape composition and the ways people use, interact with and relate to landscapes.

In Europe, the European Landscape Convention (ELC) promotes the protection and management of landscapes (Council of Europe, 2000). Signatory countries are obliged to analyse landscape characteristics and the processes transforming them, as well as assessing the values people assign to landscapes (ELC articles 6C 1a and 1b). But how can we assess the complexity of landscape and monitor change? One commonly applied strategy is to develop methods to quantify and monitor landscape change and the effects of policies at local, regional and national levels. These methods can take a range of forms (Kienast et al., 2019) incorporating, for example, indicator-based approaches measuring and combining landscape properties as indicators of associated landscape patterns and processes or using LCA to descriptively document the state and change of landscapes (see the next vignette for more detail on this).

One advantage of landscape change indicators is that simple numerical measures allow quantification of change, which is welcomed by policy-makers seeking baselines to assess the effect of policies. However, indicator-based monitoring approaches often focus on more physical aspects of landscape, such as changes in land cover and land use (Peano and Cassatella, 2011). Examples include the amount of agricultural land converted to new settlement areas, change in forested area or length of revitalised rivers. However, simply measuring changes in land cover and land use tells us little about how societies and individuals perceive and experience this change, or about their relationship with landscape and its value to a society. Within indicator-based frameworks, perceptions and value-related questions can be integrated through tools such as surveys, questionnaires and public participation Geographic Information Systems (GIS). Participants, typically residents, can give views about landscape value, for example, with respect to aesthetic properties, the activities they undertake, and feelings and meanings they attach to landscapes. Such assessments are based on notions of landscape as places imbued with meaning, and link to an extensive literature and debate on concepts such as place and sense of place (Tuan, 1977; Hirsch and O'Hanlon, 1995; Cresswell, 2013; Massey, 2013).

Monitoring including people's connection to landscapes, as well as meanings associated to landscapes along measures of physical landscape change thus attempts to bridge the gap between landscape constituted by bio-physical landscape elements such as mountains, rivers and meadows and the cultural and emotional notions people associate with these elements and the landscapes they are part of.

Flurina Wartmann
Department of Geography & Environment,
University of Aberdeen, UK

Indicator-based approaches such as those described by Flurina often use rich spatial data, for example, in the form of land cover and land use maps. Historical Landscape Characterisation, introduced in the next vignette by Graham Fairclough, also uses spatially explicit data (e.g., in the form of aerial photographs), but places much more emphasis on expert interpretation. LCA is of particular interest, because narrative (in the form of, for example, the literature written about a region) flows directly into its production. As Graham makes clear, narratives, past, present and future lie at the core of LCA as a method.

Landscape character

In Britain and a growing number of other countries (Fairclough, Sarlöv Herlin, and Swanwick, 2018), the assessment or description of landscape character is a standard practice for the understanding and management of landscape in a managerial, conservation and environmentalist context. 'Landscape', it should be emphasised, is in this field treated as a broadly defined concept, currently best summed-up by the definition of the European Landscape Convention that landscape is 'an area perceived by people whose character is the result of the action and interaction of natural and/or human factors' (Council of Europe, 2000). In this definition, perception is much more than 'visual' but embraces all ways of sensing, feeling, remembering and understanding, and landscape means many different things to different people. It is a highly fluid, ambiguous and potentially contested notion whose study crosses many disciplinary boundaries and reaches out towards a transdisciplinary response to significant social, environmental and cultural challenges (Bloemers et al., 2010).

In the British situation two interrelated approaches exist. Both utilise the concept of 'character' and both share common aims targeted on landscape management and heritage conservation. They adopt different approaches to analysis and process, but both are generalising procedures which aim to grasp the overall character of landscape by capturing human and cultural perceptions across relatively large areas. They are of necessity selective in which aspects of landscape are privileged and thus susceptible to absorbing unacknowledged biases and assumptions.

Historic Landscape Characterisation (HLC) was developed from the mid-1990s as a complement to preexisting methods of LCA (Fairclough and Herring, 2016). The LCA approach sought to be holistic in its understanding of landscape, but for a variety of reasons (not least non-availability of spatially comprehensive accessible data) tended to understate the historic dimensions in the present landscape. LCA was also limited in its spatial orientation, offering primarily area-based written descriptions, which makes its use in spatially oriented managerial contexts more difficult. In contrast, therefore, HLC developed spatially based (within GIS) and historically focused methods to sit alongside LCA's mainly text-based, largely visual appreciation of landscape.

As hinted above, 'landscape' is a fuzzy concept in the first place, but this is exacerbated when studying the past because information is almost by definition incomplete and uncertain, not always spatially located and often not place-specific (but instead derived by extrapolation from knowledge gained in comparable locations elsewhere). HLC interestingly sought to deal with this by relying on a distinctively unfuzzy framework, that of GIS, to contain the HLC. GIS enabled the construction of a continuous seamless coverage of spatial polygons, which lends a spurious objectivity to the process, particularly when those polygons are used to contain a mass of information that is better characterised as 'interpretation' rather than 'data'. This created tension with and over-simplified the 'data' – strictly speaking in fact an interpretation of data – that HLC used. HLC's primary source (following an archaeologist's approach to material culture) is the physical manifestation of the landscape itself, interpreted through its morphology, constituents and evolution within an imposed classification of historically informed types. The analysis uses distanced images such as aerial and satellite photographs and images and maps modern and historic which enable a detailed, fine grain. The end result is an interrogatable database[2] of spatially related information that awaits future interpretation and use in the practical contexts for which it was designed, notably landscape and heritage management and spatial planning. HLC projects produce typological or chronological analyses and only but less commonly

---

[2] Downloadable examples can be found at https://archaeologydataservice.ac.uk/archives/view/HLC/

extensive written descriptions of places or localities, and the scope for language analysis is probably limited.

LCA projects on the other hand offer primarily text-based descriptive overviews. These sit within a simpler territorial framework of very large areas, generally perceived from an on-the-ground, horizontal perspective. Narratives of various types are at the centre of the LCA method (even though in some applications, such as the English national character areas, they have since been complemented by databases, for instance, ecological data, selected to match the LCA areas). LCA is not rooted in a detailed spatiality, sometimes not even within a GIS, although the territory involved – the study area – is initially subdivided in an intuitive manner into a set of continuous uniquely defined 'landscape character areas'. While scales can vary, these areas tend to be relatively large, usually well above the level of what is commonly meant by 'place'. As such, they can in a relatively straightforward way become containers for descriptions and information which therefore allows the production of coherent, aspirationally definitive narratives that are then susceptible to automated comparison and analysis. The LCA texts are in essence environmental narratives. LCA also comes close (to a certain extent in contrast to HLC with its focus on human agency in landscape and its understanding of the long-term inevitability, past and future, of physical change) to providing environment-alist narratives based on preservationist aspirations, anxieties against change and often a romanticised nostalgia. Furthermore, because in disciplinary terms LCA has historical antecedents in the field of topographical writing and of what used to be called 'natural history', it often takes raw material from past literary (and for that matter visually artistic) descriptions of an area, especially in drawing from past narratives a sense of value of cultural association. It thus both channels past narratives and constructs present-future narratives. LCA has also in modified forms been taken up in other countries, and therefore analysis can be made of the impact of national cultural attitudes (e.g., different cultural approaches to modernity, change and preservation, or the attitudes to and definitions of landscape and heritage values) and of linguistic contexts (e.g., the deep differences between the meanings and social implications of 'landscape' and 'paysage').

Graham Fairclough
School of History, Classics and Archaeology,
Newcastle University, UK

In these first five vignettes, we have seen different ways in which narrative can be used to explore and understand the environment. Katrín and Sarah both introduced specific examples of the use of narrative to explore change in very different – (here: Icelandic rural and urban American) settings. Karen delved into the history of particular words and concepts, and explores how a concept bridging urban and rural landscapes – the park – emerged as an idea. Flurina and Graham introduced ways in which landscape, another very specific concept, is monitored and discussed how narrative can and could feed into this process. But not only words have origins, narrative itself also has a variety of forms, and our last vignette, from digital humanities scholar Gabriel Viehhauser, introduces ways in which space can be used as a way of understanding, and starting to analyse, narrative.

(Digital) narratology and space

As is true for probably every other cultural studies discipline, the spatial turn of the 1990s also affected literary studies, the field that would probably claim to have the most genuine expertise in describing narratives. The turn led to a plethora of studies on spatial constellations in literature, describing them in relation to different historical or cultural contexts or formal aspects like genres. However, it often applied a very metaphorical concept of space that sometimes risks losing touch with the concrete spatial settings of a literary work.

This interest in the category of space could draw on work that, long before the spatial turn, pointed out the importance of a semantic dimension of spatial constellations. In doing so, this work made us aware that space could be more than just an invariant physical setting, but should be seen in relation to the (human) beings that move in spaces. Some of the classical studies include Bakhtin's works on the so-called chronotopoi, spatial-temporal patterns that can be related to different historic genres (Bakhtin, 1981), or the model of spatial semantics developed by Lotman connecting plot structures, characters and space and their significance in telling a story (Lotman, 1977).

Interestingly, the very basic question, as to how space is constructed in narratives at all, received far less attention than those considerations on changing spatial constellations and their meaning. This is surprising, since inquiries investigating, for example, where a story is set and how this fact is expressed in words, underlie all other endeavors to exploit the semantic dimensions of space.

I would argue that with the advent of digital methods in literary studies and the possibility to perform distant reading of
huge amounts of texts with the help of computers these questions deserve renewed attention, since many research problems
related to space might be amenable to such an approach. For
example, the examination of chronotopoi in different historical
genre-constellations performed by Bakhtin and his successors is
necessarily often based on a selective text base, since genres can
encompass more works than a single researcher can read. Given
a clear concept on how space is constructed and how this construction can be identified with the help of computers, those studies could easily be enriched with information drawn from larger
corpora. To give another example, studies from the field of literary geography (e.g., as proposed by Moretti (1999) or Piatti (2008;
2009) that strive to trace the settings of literary texts in geographical maps and to show which places are described more or less often
in the literature of a given time or place could also support their
evidence by exploiting more texts, if this could be analysed and
mapped automatically.

Unfortunately however, the question as to how space is evoked
in a text does not have an easy answer. In narratology, the sub-
discipline that deals with the construction of narratives, there
are far fewer studies on space than on other basic categories like
time, character or plot. This is probably partly due to the fact
that the creation of narrative space relies even more on implicit
meanings than those other categories: space is a pre-condition
of every narrative, but exactly because of this it does not have
to be explicitly mentioned or described every time it is evoked.
Characters move through and act in space, thus constituting space
implicitly. Practically every object that appears in a narrative can
be situated in space and therefore has potential spatial dimensions. And furthermore, non-spatial elements like professions or
activities could point towards spatial constellations without even
mentioning them: if a character, for example, wakes up in the
morning, it is likely that she is lying somewhere, implying a bed
or some other object that is supplemented by the reader's mind
according to their imagination of the story world. If the first thing
that she sees in the morning is a nurse, it is likely that she is located
in a hospital or a similar facility.[3]

---

[3] Which notion exactly springs up in the mind of the reader can be described with
the 'principle of minimal departure', formulated by Ryan: the reader will tend to
imagine a story world that resembles her own in as many aspects as possible, as
long as the text does not indicate otherwise (Ryan, 1980).

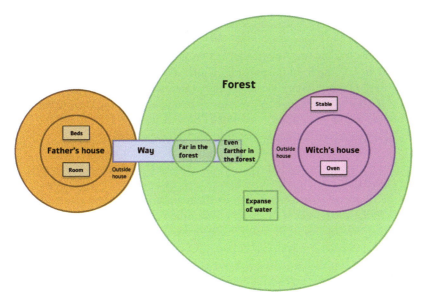

Figure 2.4: Spatial frames in the fairy tale 'Hansel and Gretel'.

In other words, there is a huge gap between the words on the textual surface that can create a spatial setting and the mental model of that space that we develop while reading a text. This notion has astutely been described by Ruth Ronen in her concept of 'spatial frames' (Ronen, 1986). 'Spatial frames' are mental concepts that aggregate the verbal hints in a text which point towards a spatial setting (like in the example above the nurse pointed towards the spatial frame 'hospital').

In Figure 2.4, I tried to trace down a diagram of spatial frames for a rather basic and famous fairy tale 'Hansel and Gretel'. The orange circle on the left signifies Hansel and Gretel parents' house, from which they are sent twice to the forest (green circle) by their evil step-mother. Whereas the children manage to find their way back the first time, they get lost the second time, ending up in the witch's house (purple circle, a frame inserted in the larger frame of the forest), where they are caught, but finally overcome the witch.

The crucial question for a digitally informed narratology would be: Is there a way back from the space markers on the textual surface (that can be detected by the computer) to the underlying spatial frames (which could be understood as some sort of latent concepts underlying the text)? A digital aggregation of spatial markers into spatial frames would not only build the base for a spatial distant reading of texts, but also support concepts such as

Lotman's spatial semantics mentioned above. Lotman claims that a story only has an eventful plot (a 'sujet'), if a character of the story transgresses the border between two semantically different spheres. While in 'Hansel and Gretel' the parents and the witch rest in their own spheres for the whole narrative, the two children move from one space to another, and because the two places are linked to different semantics ('normal world' vs. 'fairy world', 'home' vs. 'outland'), in doing so, they trigger an event.

Once the problem of the aggregation of space is solved, the distinct semantic load of the frames could be traced back by digital means.

Figure 2.5 shows the most distinctive words of the passages set in the witch house (which here I identified manually) compared to the other parts of the story.[4] The words are not explicitly related to a fairy-tale world, but reveal a different semantic connotation of the frame that perhaps is not obvious at first glance. It is conspicuous that female characters play a major role in the sphere of the witch house, where the orthodoxy of gender relations seems to be turned upside down: whereas in the father's house and in the woods Hansel is active and his sister remains passive and scared, Gretel takes command in the witch's house, where she is the one that burns the witch and actively intervenes, whereas her brother is forced to remain passive, because he is locked into the stable. Examples like this show, that a digital narratology of space could be a promising tool for tracing spatial constellations in narratives.

Gabriel Viehhauser
Institute of Literary Studies, University of Stuttgart

Our last vignette discussed a very different kind of narrative, that of a fairy tale. Nonetheless, this fairy tale is also set in an environment with a history as described by Karen Jones, and as Katrín Anna Lund showed, mythical beings also have their place to play in understanding modern landscapes. Gabriel shows how what he calls digital narratology can allow us to explore the relationship between physical settings and the characters of the fairy tale. His vignette hints at the potential, but also the need for methods which allow exploration of texts in novel ways, in this case related to the locations of the participants of the narrative in space. Taken together, the six narratives show a range of perspectives from which narratives can be interpreted and understood.

---

[4] For the comparison I used log-likelihood-measurement, performed with the help of the R Quanteda package (Benoit et al., 2018).

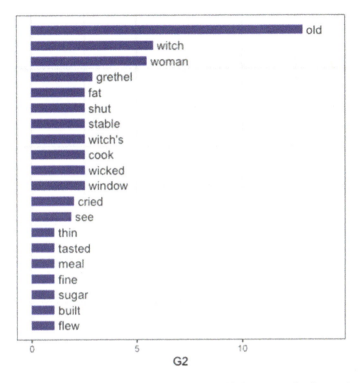

Figure 2.5: Distinctive words related to the witch's house in the fairy tale 'Hansel and Gretel'.

To unlock these narratives, by allowing their analysis and interpretation at scale, we need to start to consider if, and how, computational approaches can help. Doing so requires not only methods, but also digital collections of documents. In Chapter 3 we illustrate, through an example, ways in which we can start to explore text computationally.

References

Bakhtin, Mikhail (1981). "Forms of time and of the chronotope in the novel". In: *The dialogic imagination. Dialogic imagination: Four essays by M. M. Bakhtin.* Ed. by Michael Holquist. Austin: University of Texas Press.

Bender, Barbara (2002). "Time and landscape". In: *Current Anthropology* 43, S103–S112. DOI: 10.1086/339561.

Benoit, Kenneth, Kohei Watanabe, Haiyan Wang, Paul Nulty, Adam Obeng, Stefan Müller, and Akitaka Matsuo (2018). "Quanteda: An R package for the quantitative analysis of textual data". In: *Journal of Open Source Software* 3.30, p. 774. DOI: 10.21105/joss.00774.

Bloemers, Tom, Stephen Daniels, Graham Fairclough, Bas Pedroli, and Richard
Stiles (2010). *Landscape in a changing world - Bridging divides, integrating
disciplines, serving society.* Tech. rep. Strasbourg/Brussels.

Council of Europe (2000). "European landscape convention". In: *Report and
Convention Florence* ETS No. 17.176, p. 8. URL: http://conventions.coe.int/
Treaty/en/Treaties/Html/176.htm.

Cresswell, Tim (2013). *Place: A short introduction.* Hoboken: John Wiley &
Sons.

Cronon, William (1992). "A place for stories: Nature, history, and narrative". In:
The Journal of American History 78.4, pp. 1347–1376. DOI: 10.2307/2079346.

— ed. (1995). *Uncommon ground: Toward reinventing nature.* New York: W. W.
Norton.

Dovers, Stephen (1994). "Australian environmental history: Introduction,
review and principles". In: *Australian environmental history: Essays and
cases.* Oxford: Oxford University Press.

Fairclough, Graham and Pete Herring (2016). "Lens, mirror, window: Interactions between historic landscape characterisation and landscape character
assessment". In: *Landscape Research* 41.2, pp. 186–198. ISSN: 14699710. DOI:
10.1080/01426397.2015.1135318.

Fairclough, Graham, Ingrid Sarlöv Herlin, and Carys Swanwick, eds. (2018).
Routledge handbook of landscape character assessment. London and New
York: Routledge. DOI: 10.4324/9781315753423.

Fell, Mary (1984). "The prophecy". In: *The persistence of memory.* New York:
Random House.

Hibbett, Maia (2019). *What Happens When Gentrification Comes to a Postindustrial City?* https://www.thenation.com/article/gentrification-worcester-
urban-development-homelessness/. [Online; accessed 17-October-2019].

Hirsch, Eric and Michael O'Hanlon (1995). *The anthropology of landscape: Perspectives on place and space.* Oxford: Oxford University Press.

Kienast, Felix, Flurina Wartmann, Alice Zaugg, and Marcel Hunziker (2019). "A
review of integrated approaches for landscape monitoring. Report prepared
in the framework of the Work Program of the Council of Europe for the
implementation of the European Landscape Convention. Report no. CEP-
CDCPP (2019) 7E". Strasbourg.

Lotman, Jurij (1977). *The structure of the artistic text.* 7. Ann Arbor: University
of Michigan/Michigan Slavic.

Lund, Katrín Anna (2015). "Just like magic: Activating landscape of witchcraft
and sorcery in rural tourism, Iceland". In: *Changing world religion map:
Sacred places, identities, practices and politics.* Ed. by S. Brunn. New York:
Springer, pp. 767–782. DOI: 10.1007/978-94-017-9376-6_38.

Lund, Katrín Anna and Karl Benediktsson (2010). "Introduction: Starting a
conversation with landscape". In: *Conversations with landscape.* Ed. by Karl
Benediktsson and Katrín Anna Lund. Farnham: Ashgate, pp. 767–782.

Lund, Katrín Anna and Gunnar Thór Jóhannesson (2014). "Moving places: Multiple temporalities of a peripheral tourism destination". In: *Scandinavian Journal of Hospitality and Tourism* 14.4, pp. 441–459. ISSN: 15022269. DOI: 10.1080/15022250.2014.967996.

— (2016). "Earthly substances and narrative encounters: Poetics of making a tourism destination". In: *Cultural Geographies* 23.4, pp. 653–669. ISSN: 14744740. DOI: 10.1177/1474474016638041.

Massey, Doreen (2013). *Space, place and gender*. Hoboken: John Wiley & Sons.

Merleau-Ponty, M. (2002). *Phenomenology of perception*. London: Routledge. DOI: 10.4324/9780203994610.

Moretti, Franco (1999). *Atlas of the European novel, 1800-1900*. London: verso.

Morgan, Ruth A. (2013). "Histories for an uncertain future: Environmental history and climate change". In: *Australian Historical Studies* 44.3, pp. 350–360. DOI: 10.1080/1031461X.2013.817456.

Olmsted, Frederick Law (1881). *Mount Royal, Montreal*. New York: Putnam's Sons.

Peano, Attilia and Claudia Cassatella (2011). "Landscape assessment and monitoring". In: *Landscape indicators*. Springer, pp. 1–14. DOI: 10.1007/978-94-007-0366-7_1.

Piatti, Barbara (2008). "Die geographie der literatur". In: *Schauplätze, Handlungsräume, Raumphantasien*.

Piatti, Barbara, Hans Rudolf Bär, Anne-Kathrin Reuschel, Lorenz Hurni, and William Cartwright (2009). "Mapping literature: Towards a geography of fiction". In: *Cartography and art*. Berlin: Springer, pp. 179–194. DOI: 10.1007/978-3-540-68569-2_15.

Quinn, Tom (2016). *Canal Digging: Is Worcester's Canal District the City's Real Downtown Core?* https://www.worcestermag.com/2016/06/30/canal-digging-worcesters-canal-district-citys-real-downtown-core. [Online; accessed 17-October-2019].

Ronen, Ruth (1986). "Space in fiction". In: *Poetics Today* 7.3, pp. 421–438. DOI: 10.2307/1772504.

Rose, Mitch and John W. Wylie (2006). "Guest editorial: Animating landscape". In: *Environment and Planning D: Society and Space* 24.4, pp. 475–479. ISSN: 02637758. DOI: 10.1068/d2404ed.

Rosenzweig, Roy (1983). *Eight hours for what we will: Workers and leisure in an industrial city 1870-1920*. Cambridge: Cambridge University Press.

Ryan, Marie-Laure (1980). "Fiction, non-factuals, and the principle of minimal departure". In: *Poetics* 9.4, pp. 403–422. DOI: 10.1016/0304-422X(80)90030-3.

Schacter, Aaron (2018). *Forget Oakland or Hoboken, Worcester, Mass. is the New It Town*. https://www.npr.org/2018/10/23/658263218/forget-oakland-or-hoboken-worcester-mass-is-the-new-it-town. [Online; accessed 05-May-2020].

Smith, Neil (1996). *The New Urban Frontier: Gentrification and the Revanchist City*. New York: Routledge.

Tuan, Yi-Fu (1977). *Space and place: The perspective of experience*. Minneapolis: University of Minnesota Press.

Williams, Raymond (1983). *Keywords: A vocabulary of culture and society*. London: Flamingo.

Worster, Donald (1988). *The ends of the earth: Perspectives on modern environmental history*. Cambridge: Cambridge University Press. DOI: 10.1017/CBO9781139173599.

CHAPTER 3

Analysing Environmental Narratives Computationally

Ross S. Purves

Department of Geography; URPP Language and Space, University of Zurich, Switzerland

Olga Koblet

Department of Geography, University of Zurich, Switzerland

Benjamin Adams

Department of Computer Science and Software Engineering, University of Canterbury, Christchurch, New Zealand

This book's origins lie in a desire to showcase, through an interdisciplinary approach, the potential for computational methods in analysing text that describes the environment. Our argument is that these computational methods need not be complex, but rather that through a combination of well-designed research questions, appropriate text collections, a sensible choice of methods and careful interpretation, we can gain new and useful insights. In this chapter we work step-by-step through a set of basic building blocks to illustrate how environmental narratives can be computationally analysed. To make our arguments clear and concrete, we divide the material that follows into two parts. Each section starts with a general overview, introducing key concepts, literature

How to cite this book chapter:
Purves, Ross S., Olga Koblet, and Benjamin Adams (2022). "Analysing Environmental Narratives Computationally." In: *Unlocking Environmental Narratives: Towards Understanding Human Environment Interactions through Computational Text Analysis*. Ed. by Ross S. Purves, Olga Koblet, and Benjamin Adams. London: Ubiquity Press, pp. 43–84. DOI: https://doi.org/10.5334/bcs.c. License: CC-BY 4.0

and methods in a general sense. Accompanying this material are worked examples, where we show how the methods discussed can be used in practice to explore one particular form of environmental narrative – discussions of the role of the Forestry Commission in the UK since its formation in 1919.

3.1   Narrative Forms

We are all familiar with different ways of writing. Good recipes are well ordered, telling us exactly how to perform each step on the way to the perfect chocolate cake, and leaving no ingredients to our imagination. Political texts carefully argue, picking out evidence backing a particular position and applying thoughtfully chosen rhetoric, to persuade us of the merits of voting for a proposition. Travel guides emphasise and richly describe particular events or places, usually with the aim of informing our visit to the same location. There are many different ways of characterising such texts, focusing, for example, on narrative form or the genre of writing. These different forms matter, because understanding how they work is important to the ways in which we interpret their contents. Rather than delve deeply into literary theory here, we would like to introduce a few important concepts which can inform how we approach computationally analysing a text.

The first of these is the **hermeneutic circle** (Martin, 1972; Boell and Cecez-Kecmanovic, 2010). Simply put, this means that our understanding of a complete text is based on our reading of its individual parts, and our understanding of the parts is based on our reading of the whole. The key idea here is that to properly understand a text we need to both interpret it as a complete work *and* explore the individual parts making up the text, hence the interpretative circle involves a continuous back and forth. At its most reductionist level, the hermeneutic circle implies that we cannot know the meanings of individual words without seeing them in context. So, given a set of parts (words) {man, dog, bit, the} we do not know whether the man bit the dog, or the dog bit the man. To interpret the sentence (the whole) we need to understand the parts and know that "man" and "dog" are objects, and "bit" is an action. Interestingly, at least in English, we can discard "the" without any change to the meaning, which is conveyed by context (men and dogs can bite) and word order.[1] For those of us working computationally, extending these ideas to longer texts requires a realisation that interpreting a text at the word- and sentence-level is not enough to fully understand it, and equally importantly, to realise that external context (e.g., the political position of an author) might also be required to understand a text.

---

[1] The role of contextual factors in understanding the meaning of a text is also studied in pragmatics where not only linguistic context but also situational context (e.g., the writer's intention, who the audience is, what the social environment is, etc.) plays a role (Green, 1996).

This importance of context and its influence of our interpretation of a text is lucidly demonstrated by the short discussions of the "Glencoe Road" text in the introduction. Each interpretation was influenced not only by the disciplinary backgrounds of the discussants, but also by their geographical knowledge (e.g., whether they were personally familiar with the region) and their general knowledge of the UK, its politics and debates.

Central to our understanding and analysis of environmental narratives are the ways in which we extract and analyse references to space and place. Doing so naively, and simply treating locations as a reality in which a narrative takes place would be to ignore much of what we know about the use of metaphor and narrative in language. Thus, the space in which a literary text takes place is not only to be interpreted literally, but also metaphorically and culturally (Lotman, 1990; Lakoff and Johnson, 2008). For example, high places are generally associated with success and well-being, and low places with failure and depression.[2] These metaphors pervade written language, and are so effective that most of us do not notice them until we encounter their (unfamiliar) use in a non-native language. Dealing with such metaphorical uses of language is essential if we wish to computationally analyse text, as otherwise a computer has no way of distinguishing between, for example, fictive motion (we followed the ridge) and real motion (we followed the car) (Egorova, Tenbrink, and Purves, 2018). The semiotician, Yuri Lotman, pointed out a second, obvious, but also often neglected point. The space described in a (literary) text is an abstraction of (some) reality, not a copy. This abstraction is influenced by both explicit choices (e.g., those of a nature writer to emphasise sounds and sights in a landscape), but also less deliberate choices, influenced by, for example, culture, background and language.

Understanding, or at least acknowledging, narrative form is important for computational analysis, since it profoundly influences both the questions that it is reasonable to ask of a corpus and the results that a computational analysis can produce. These questions can take multiple forms, but it is important to consider them before commencing an analysis. We emphasise here the importance of formulating guiding questions before starting work and, where appropriate, **hypotheses.**

Hypothesis

One reading of the text on the "Glencoe Road" in Chapter 1, that provided by Graham Fairclough, focused on the political nature of the argument presented in the article. Fairclough notes the importance and value of individual voices, specifically that of the elite in

---

[2] The pervasive use of spatial metaphors in language (e.g., "rising up the ranks", "going through a rough patch") means that much spatial language in a text – even within texts that we would characterise as environmental narratives – might have no relationship whatsoever to environmental spaces.

the form of Sir John Stirling Maxwell. He identifies a metaphorical spatial disconnection between the well-connected and privileged elite and a region on its periphery. Temporally, he positions the article with respect to the importance of two movements - one connected with modernisation and progress through the motor car, and the other with the start of various movements connected with countryside conservation.

Fairclough's reading of the text contained a footnote which provides the starting point for the exemplary analysis carried out in this chapter. He notes that Maxwell was also "Chair of the Forestry Commission from 1929 to 1932, an organisation that some argue has had a less benevolent impact on the British landscape."

The Forestry Commission was formed in 1919, as a response to the depletion of Britain's forests during the First World War. Its original role was concerned with forestry as a means of producing timber, but over time this has changed to one more and more influenced by debates concerned with landscape beauty, biodiversity and recreational use of forests.

To explore the role of the Forestry Commission over time as recorded in text we turn to Hansard, edited transcripts of the proceedings of the British Parliament, which documents in great detail the activities of the House of Commons and the House of Lords. Inspired by Fairclough's footnote, we set out to explore how the Forestry Commission has been discussed in the British Parliament since its inception in 1919, with a focus on the perceived impact of the Forestry Commission on landscape. As a second, more contemporary source reflecting the views of those active in the countryside, we explored descriptions from Geograph[3], a crowdsourced collection of more than six million images and texts describing 1 km grid squares across Great Britain.

We hypothesise that discussion of the Forestry Commission in parliament will focus on the perceived negative impact of forestry, and the Commission's activities, on landscape. In Geograph we expect to see similar patterns, but with a more direct concentration on the visual impact of Forestry Commission activities in the landscape, since in this case our collection consists of texts describing photographs.

3.2   Building a Corpus

The first step in working with texts is compiling a large, systematic text collection, commonly known as a corpus (corpora in plural). The meaning of

---

[3]   https://www.geograph.org.uk/

large and *systematic* depends on the research questions we want to answer with this corpus. Creating corpora can be traced back to at least the 1950s, when literary scholars started compiling systematic text collections of, for example, the complete works of one author or of a variety of authors covering the same **time period** (Pustejovsky and Stubbs, 2012). From the 1960s, influenced by the needs of corpus linguistics, corpora capturing usage of American and British English started to be created, in the form of, for example, the Brown Corpus and Lancaster-Oslo-Bergen corpora, respectively. Since both set out to reflect general usage of *language*, texts in these corpora include both written and transcribed spoken texts as well as different **genres** and **domains**. Genre describes general characteristics of texts, for example, broadcast news, as in the corpus of regional newspapers from the UK compiled by Lansdall-Welfare et al. (2017). Domain refers to main subject of the texts, which might influence word sense (e.g., *bank* in financial texts and *bank* in river renaturalisation texts) and the specific vocabulary used in a corpus (Augenstein, Derczynski, and Bontcheva, 2017). Examples of domain-specific corpora include a multilingual corpus of mountaineering texts called Text + Berg (Sennrich et al., 2009), the Nottingham Corpus of Geospatial language (Stock et al., 2013), or a corpus of reports covering 18 years of international climate negotiations (Venturini et al., 2014). Comparison of a domain-specific corpus (e.g., a corpus of travel reports) to a general one (e.g., the British National Corpus) can reveal which words or phrases are distinct or appear statistically significantly more or less often than in a domain-specific corpus (Kilgarriff, 2001) as we will see later in Section 3.4.1.

An additional important property of a corpus in an environmental context is its **spatial coverage**, that is to say the distribution of places described in a corpus. For example, the Corpus of Lake District Writing (CLDW) (Butler et al., 2017) covers the English Lake District, while the Palimpsest corpus (Alex et al., 2016) is a collection of fictional and historical texts related to Edinburgh, Scotland.

The massive growth in the availability of digitised texts has greatly increased the range and variety of resources from which corpora can be created. These include very large collections of digitised books, the web itself and more specific collections such as newspaper archives, collections of legal documents, scientific articles or some of the corpora described above. Examples of such collections include digitised books hosted by Google Books[4] or Project Gutenberg[5], crawls of the web such as the Common Crawl[6], historical newspaper archives as provided by the Chronicling America project[7] and archival records such as Hansard recording UK parliamentary debates[8]. However,

---

[4] https://books.google.com/
[5] https://www.gutenberg.org/
[6] https://commoncrawl.org/
[7] https://chroniclingamerica.loc.gov/
[8] https://api.parliament.uk/historic-hansard/index.html

such resources are often too general to allow exploration of specific research questions, and the first step towards such an analysis is defining what properties the required corpus should have (e.g., language, spatial coverage, domain, etc.). Having defined such properties, we can extract potentially relevant documents by, amongst other possibilities, restricting ourselves to a particular genre (e.g., natural history writing in the Country Diaries of the Guardian[9]), using search terms relating to concepts in a given domain (e.g., 'glacier', 'ice', and 'mountain') or compiling lists of place names in the region of interest (e.g., 'Loch Lomond', 'Balloch', etc.) to extract documents referring to specific locations (Davies, 2013).

How can we work with sources such as those described above? Many are too large to simply copy and process in their totality locally. Often, the publishers of such data make them available through an **application programming interface (API)**. An API allows us to query a system with a defined request, and in return receive a response message. Such requests are usually returned as structured data in formats such as Extensible Markup Language (XML) or JavaScript Object Notation (JSON). These are typically hierarchical and allow data to be explored using attribute-value pairs. Thus, in the JSON fragment of a parliamentary debate from Hansard shown in Listing 3.1, "name" is an attribute with the value "Rachael Maskell". Since 'attribute-value' pairs are stored hierarchically, "name", "party" (political affiliation), and "constituency" (the geographic area represented by a Member of Parliament in the UK) are all accessible as child attributes of the parent attribute "speaker", in this case associated with "hdate", a date, presumably on which the question referred to by "body" was posed.

```
1  {
2      "body": "Bishop Wood is being used for shooting--land
           leased by the Church Commissioners to the Forestry
           Commission. Blood sports in exchange for blood money
           for the Church of England. What steps have the
           Church Commissioners taken to ban blood sports
           across their estate?",
3      "hdate": "2019-03-28",
4      "speaker": {
5          "name": "Rachael Maskell",
6          "party": "Labour/Co-operative",
7          "constituency": "York Central"
8      }
9  },
```

Listing 3.1: A JSON fragment from a UK parliamentary debate retrieved using the API provided by TheyWorkForYou (https://www.theyworkforyou.com/api/).

---

[9] https://www.theguardian.com/environment/series/country-diary

Figure 3.1: Example of a webpage with its HTML structure.
Source: https://www.theyworkforyou.com/debates/?id=2019-03-28b.
545.8#g545.11

Using an API and search terms to extract relevant documents from a given collection is the preferred approach where possible since it makes work easier to reproduce, and APIs typically also provide, firstly, metadata expanding on the semantics of attributes and, secondly, explicit terms and conditions with respect to the use of data (Section 3.3). However, many sources have been digitised and made accessible online without the implementation of APIs allowing direct querying of documents. For example, the content retrieved using an API query in Listing 3.1 is also accessible as a web page (shown on the left of Figure 3.1). The raw HTML used to render this web page is shown on the right of Figure 3.1. Inspecting this source, it is possible to identify classes we are particularly interested in, for example, the question ("body" above) is stored in the class "debate-speech__content" and the speaker ("name") in class "debate-speech__speaker__name". We can access these classes and extract the information they contain using a web scraper. Different programming languages have libraries specifically created for this task, for example, Scrapy[10] and Beautiful Soup[11] in Python, rvest[12] in R, and JSoup[13] in Java. This approach allows us to extract the content of a single web page or a family of web pages with the same or very similar structures. It is therefore well-suited to automatically extracting content from consistently structured web pages such as Wikipedia or content that follows a well-defined template, for example, reports produced by government agencies.[14] In doing so, it is important to consider any copyright and ethical considerations (Section 3.3).

---

[10] https://scrapy.org/
[11] https://pypi.org/project/beautifulsoup4/
[12] https://cran.r-project.org/web/packages/rvest/
[13] https://jsoup.org/
[14] Often websites like Wikipedia prefer that you download their data as packaged database dumps (see https://dumps.wikimedia.org), rather than via web scraping, which can put strain on their web servers and slow down the website for casual users. It is important to read the terms of use of any website that you are planning to scrape and avoid pages that are explicitly banned from automated scraping.

When we explore environmental narratives using online sources, we are often interested in building a corpus of documents describing a particular theme and/or region. If we aim to explore different domains, genres or perspectives, we may also want to analyse different sources, and scraping a single website is no longer sufficient. Here, we can take a similar approach to that described above using an API to search a specific collection or using a general web search API to search the web as a whole. The first step in such work is compiling a list of search terms (e.g., place names in a given region and word associated with a particular topic). Search engine APIs usually return uniform resource locators (URLs) rather than full document text in a first step, and the content of these URLs can be extracted using the scraping methods described above (with the important difference that the structure of individual web pages is likely to vary widely). Since irrelevant documents are also highly likely to be returned (e.g., hotel rooms named after places), filtering steps are also necessary when building a corpus using this approach. The BootCaT tool[15] makes collecting documents using search engine APIs possible without programming experience and although it is limited to 100 URLs, is very useful in exploring and testing queries and ideas.

Corpora

Since we wished to explore how the Forestry Commission was discussed in the UK Parliament, we first looked for online sources of debates. Hansard is the official written record of British parliamentary proceedings, is available and searchable online at https://hansard.parliament.uk/ and has been used to explore for example infrastructure in the British Empire (Guldi, 2019). To search the collection for discussions about the Forestry Commission programmatically, we took advantage of a third-party API implemented by the UK-based organisation mySociety (https://www.mysociety.org/). The API (https://www.theyworkforyou.com/) allows us to query for information on a variety of dimensions, including debates held by the upper (House of Lords) and lower chambers (House of Commons) of parliament. In an initial exploration we searched Hansard for all mentions of 'Forestry Commission', returning the transcripts of debate, their dates, speakers and whether the debate took place in the Commons or Lords. For the House of Commons, the API returned 1985 documents, contrasting sharply with the 190 returned from the House of Lords. We quickly established that while the documents from the Commons dated back to 1919

---

[15] https://bootcat.dipintra.it/

(the year the Forestry Commission was founded), those from the Lords started only in 1999. We therefore decided to concentrate on Commons debates. In a second filtering step, we identified a large number (408) of very short documents with no speaker assigned. Our final corpus therefore contained 1577 documents from Hansard debates recorded in the House of Commons dating back to 1919.

When building a corpus, identifying appropriate sources and exploring their properties, before starting analysis is important. Doing so requires a basic knowledge of the expected properties of documents related to the theme (here, we know when the Forestry Commission was founded) and collection (we would expect broadly similar temporal periods in documents returned from the two chambers of the UK parliament).

Geograph texts can be downloaded as a single corpus, and we queried the complete database for all descriptions mentioning the Forestry Commission, identifying 3114 such texts.

3.3 Copyright and Ethics

In an era where large volumes of text are readily available online, it is all too easy to gain the impression that, quite literally, anything goes. We can, as was discussed above (Section 3.2), develop crawlers to scrape data and build corpora based on any content that is visible on the web. However, when we build such corpora we need to be able to answer two, linked, but separate questions. Firstly, are we legally allowed to use these texts in the way that we plan to? And secondly, and equally importantly, are there ethical issues that should be considered before we commence our study? It is important to understand that legal and ethical standards change in space and time. For example, copyright laws vary according to legal jurisdictions and acceptable ethical practices change over time. In what follows, we give a non-exhaustive list of issues to consider when designing an experiment, and conclude with a checklist of questions to ask before starting work.

The increased importance of reproducibility in research has brought with it a recognition of the need to provide data and code together with scientific papers reporting on research results. This welcome development allows researchers to replicate existing results, and build upon them more easily than in the past. With respect to research on text, shared datasets allow the development of common baselines (e.g., with respect to identifying locations or characterising sentiment) based on published corpora and related annotations. Furthermore, given the complexity and challenges involved in building domain-specific corpora, reusing these for other research reduces duplication of effort and allows research to more directly and comparably build upon previous work.

However, before publishing a corpus, it is important to understand the notion of **copyright**. Simply put, copyright protects the creator of a work from its reproduction without their permission. Copyright laws vary widely geographically, but typically are of long duration, often extending 50 years or more beyond the death of the creator or author. Thus, books, newspaper articles, scientific papers and images are all usually protected by copyright which requires explicit permission for the reproduction and publication of material. Copyright holders may give permission for academic use, but simply by providing access to content, copyright holders do not relinquish their rights. In some countries, the notion of fair use allows limited quoting or reproduction of content in certain contexts, with, for example, quoting from a song or a book permitted as part of a review, reportage or even parody. In the UK for instance, academics and students carrying out non-commercial research are explicitly allowed to carry out text and data mining of sources to which they already have access through, for example, subscriptions[16].

For many works, information about copyright is explicitly provided. Thus, scientific publishers and newspapers publish copyright statements, and explain how works can be licensed for further use. Typically, such licensing is complex and may involve additional fees, based for example on the number of users accessing the content. One important development is the increase in the use of explicitly permissive licences, such as Creative Commons. Here, authors give permission for their work to be reused in different ways. The most open such license is Creative Commons Zero (https://creativecommons.org/share-your-work/public-domain/cc0/), which places a work in the public domain with no restrictions whatsoever. However, much more widespread are licences which require attribution of a work, restrict commercial reuse or prohibit derivative works.

In general, collections of unstructured text used for analysis of environmental narratives can be categorised with respect to licensing in four broad categories:

- Licence allows redistribution and adaption under no, or some conditions (e.g., Wikipedia and Geograph) allowing corpora to be created directly incorporating such material. Where licences vary, then care must be taken in merging materials.
- Curated corpora created by third parties and shared under clear licence conditions (e.g., Text+Berg where licence has been negotiated with copyright owners or the Corpus of Lake District Writing which consists of historical, out-of-copyright documents).
- Licensed corpora (e.g., GeoCLEF) of newspaper articles, where redistribution is subject to restrictions and permission from the licensee or copyright holder.
- Scraped corpora of content from the web, blogs and so on where fair use may be implicitly assumed but copyright is unclear.

---

[16] https://www.gov.uk/guidance/exceptions-to-copyright

In practice, it appears that many researchers working with text build corpora with limited regard to the situation with respect to licensing. For example, the Geograph project consists of millions of images and accompanying textual descriptions located all over the British Isles. Individual contributions are licensed with a Creative Commons BY-SA licence (https://creativecommons. org/licenses/by-sa/2.0/). This licence allows copying, redistribution and transformation of the content for any use, even commercially, under two conditions. Firstly, appropriate credit to the author must be given. Secondly, any new material developed based on the source must be distributed under the same licence conditions. We identified 32 research papers which had used these Geograph data, from many different research groups. Of these, only 17 attributed the data as required in the licence, and even less made their results available under an equivalent licence.

This lack of regard for clearly communicated licensing conditions for the reuse, adaption and distribution of text brings us to the question of ethics. Ethics have traditionally been policed academically by institutional review boards, whose domain has gradually extended from medical research, through social sciences to data analysis. Geographically, the degree of ethical scrutiny of research with respect to data and methods has varied, leading to different notions of acceptable ethical practice. Zimmer (2018) set out an ethical framework in the context of big data, exploring ethics in the context of particularly extreme example, where researchers scraped the content of an online dating website, arguing that the data were in any case public. Zimmer introduces some generally accepted principles of research ethics, including minimising harm, gaining informed consent and maintaining privacy and confidentiality. To these, we add an additional idea, transparency.

When using text to explore environmental narratives, we can aggregate individual texts to allow a macroanalysis, and zoom into particular details to perform microreading. These scales are important, as they also have ethical implications relating to the three principles identified by Zimmer. Macroanalysis typically obfuscates individual contributions, reducing our ability to identify individuals. Microreading, by contrast, emphasises context when reading a text, and zooms in to individual statements.

Minimising harm implies that participants, or in our case those who create content, are not subject to any harm through being involved in research. Analysing text, at first glance, appears to be a wholly innocuous activity. However, for example, by identifying illegal or controversial opinions in text, we can potentially expose authors to harm through, for instance, legal sanctions or unwanted online pressure. Using text to explore the properties of landscape may identify regions which are worthy of protection, and thus contribute to overall public good. But if text analysis can be used to identify such regions, then conversely it also has the potential to highlight regions where protection is no longer appropriate, at least according to our sources. Removing protection from an area may have long-term negative consequences, such as a reduction in tourist visits and income. Put simply, if we analyse environmental narratives with an expectation that the information derived can be used in decision-making, then thought should be given to the potential consequences of these decisions.

If individual texts are analysed and presented for microreading, then traditional ethics would require **informed consent**, where participants are informed in advance of the benefits and risks of participation, the aims of the research, and are given the opportunity to withdraw at any time. Text analysis rarely involves informed consent, since analysis is carried out on content produced for other purposes and often without the knowledge of the creators. Thought should be given to how this can be done ethically, for instance, by linking to rather than copying content, such that where creators delete it, it is no longer used in analysis or presentation of results. In particular cases – for instance, if mapping potentially controversial statements – it may be desirable to ask contributors for informed consent before analysis and publication.

Privacy and confidentiality are not only ethical issues but also legal ones. In Europe, the General Data Protection Regulation (https://gdpr-info.eu/) sets out clear rules with respect to the processing of personal information. Ethically though, irrespective of the legal situation, we need to consider the rights of individuals to privacy by, for example, not seeking to use other data to identify individuals and giving careful consideration to whether any personal information (e.g., age or sexual orientation) should be collected without specific, informed consent.

However, these principles bring with them other challenges. If we are to maintain confidentiality, that in turn implies not attributing material to its authors – which, of course, directly contradicts the need for attribution set out above. We therefore propose an additional ethical idea, which researchers should consider, **transparency**.

Transparency means that when we build a corpus, we make clear how we did so, what licences the content had, and link to the original materials rather than storing copies. It also implies that the creators of content have access to, and can comment on the results of any research, thus building and maintaining a dialogue about the use of text in research. Transparency allows individuals whose content has been analysed the opportunity to criticise our interpretation of their material and, potentially, to 'set the record' straight or ask for their material to be removed.

For the researcher starting out with text we make the following suggestions with respect to copyright and ethics:

- Identify a range of candidate text collections for the research question under investigation.
- Research, and document, the copyright conditions under which chosen text collections are published. Consider whether fair use is applicable.

- Note any conditions under which data can be used (e.g., attribution, non-commercial, share-alike).
- Ascertain whether research requires institutional ethics review.
- If analysis only involves macroreading, ensure that results are shared appropriately and discussion is possible.
- Where microreading of corpus is important, consider how contributions can be withdrawn, harm minimised, privacy and confidentiality maintained, and a transparent dialogue enabled.

Licensing of our collections

Hansard is the public record of the UK parliament, and it is published under an Open Parliament Licence[17], a very open licence, which allows commercial and non-commercial adaptation and exploitation subject to attribution. The TheyWorkForYou project stores these data in a more structured way, (https://www.theyworkforyou.com/) allowing querying using their API. Since Hansard is a public record, and the individuals we can identify are elected representatives, there are, we judge, no ethical issues in the use of these data. However, it is worth noting that in exploring historical archives we may uncover utterances which are no longer considered acceptable, and it is important to report on such material in context.

For Geograph, the data are published under a CC BY-SA licence. This in turn requires that firstly, we acknowledge individual authors of contributions we use and quote from in our analysis. Where we analyse the corpus as a whole (e.g., looking at the frequencies of individual words) we should acknowledge the creators of the corpus in a general sense. Secondly, this licence specifies that we should allow others to use our results under the same licensing terms (so-called share alike).

3.4   Corpus Linguistics and Natural Language Processing

Computationally analysing text can take a number of forms and is referred to in different fields as **corpus linguistics**, **natural language processing** (NLP), or simply **text analysis** (Manning and Schutze, 1999; McEnery and Wilson, 2003). In this book we focus specifically on the processing of written language (text). Some areas of research on text aspire to what is referred to as general artificial

---

[17] https://www.parliament.uk/site-information/copyright-parliament/open-parliament-licence/

intelligence, investigating how computers can learn to "understand" language in the same way that people do. Although the development of such computational systems could in theory give us great insight into how people think about their environments, and great progress is currently being made, we are still far from building them (Suissa, Elmalech, and Zhitomirsky-Geffet, 2022). Natural language is often ambiguous and humans, as we have discussed above, bring a wealth of background knowledge when making sense of it. When we analyse text data we must, therefore, make many simplifying assumptions.

From a practical perspective, NLP methods today provide tools to statistically analyse written text to uncover patterns in language use, topics, sentiment and different perspectives among other things. Because these methods are computational they can be used on much larger amounts of text than humans are able to read in a limited time frame, and this is where their main power lies in helping us to understand and analyse environmental narratives. However, we must also remember the limits of the models to fully capture the many nuances of natural language that a typical human reader can easily grasp.

NLP is an extremely active area of research with thousands of new articles published each year, which push the envelope on state of the art results for various shared tasks. Increasingly these methods rely on complex deep learning models that require massive amounts of data, computational resources and energy to develop. Our goal in this section is not to provide a comprehensive summary of the more advanced techniques that are currently being developed, but rather to give an introduction to a selected suite of powerful and established methods that are well-suited for environmental narrative analysis. As we will see in the case studies described in the second part of this book, simpler, established methods can be quite effective when used appropriately and in practice are often much easier to apply.

Typically, there are two stages to any environmental narrative analysis that uses NLP. The first stage is **pre-processing** and primarily involves applying methods that translate the raw text in a corpus into a form amenable to computational analysis. Many pre-processing steps involve working with text to divide it into meaningful chunks that are obvious to a human reader. These might include identifying words, sentences, paragraphs or utterances from individual speakers. Normalisation tries to match words or sequences of words onto a single canonical form (e.g., working out that '12 pm' and 'midday', U.S. and USA, or 'Zurich' and 'Zürich' all convey the same meaning). Stemming (and a closely related technique, lemmatisation) reduces inflected words to root forms with the same aim (e.g., the stems of 'snowing' and 'snowed' are 'snow').

This stage also involves **encoding** the language in the texts as **features** as well as creating new features from the raw data. There are multiple levels of structure that humans use to make sense of natural language and thus there are multiple levels at which we might encode language as features. These levels span from individual (potentially normalised and/or stemmed) words and n-grams (sequences of words), parts of speech (e.g., adjectives, nouns) and other aspects of syntax and grammar all the way to semantics and discourse. For environmental texts we might also emphasise certain features related to the domain, for example, using lists of nouns describing natural features such as 'hill', 'mountain', 'river' and so on.

A second stage often involves analysis of features to answer questions about the language in the corpus. Two important categories of NLP analysis are classification tasks and sequencing tasks. Applications of classification-based analysis in NLP include tasks such as topic classification, document similarity, sentiment analysis and stance detection. Sequencing-based applications include building translation or summarisation systems as well as interactive systems that generate answers to queries. Although both categories have potential utility in environmental narrative research, we focus primarily on classification here as it is much more straightforward to implement with existing tools, and has great utility in understanding environmental narratives.

3.4.1   Pre-processing and encoding natural language as features

A corpus of natural language text is, at its core, simply a collection of ordered words, sometimes organised into discrete documents (possibly along with metadata about those documents, such as the author or labels). The words and occasionally the individual characters that make up a document are the basic elements used to analyse text.

In some models, the order of the words in a document is not considered. These are called **bag-of-words (BOW)** models, and are predicated on the idea that the frequency of words in a document is enough of a statistical signal for us to discover meaningful information about the corpus' content. In other words, the sequence that the words occur in, which provides humans information about grammar and much of the meaning of the text, is ignored. While on the face of it a BOW model might appear overly simplistic, it can be surprisingly effective when we are dealing with large corpora.

Examining the count of frequency of a **term** (or token) is the simplest kind of BOW analysis we can perform. We use term to refer to both individual words as well as n-grams, sequences of adjacent words. For example, 'adjacent words' is a 2-gram, also sometimes referred to as a bigram. However, simple frequency measures do not alone tell us how important a term is in a given document, since we first have to control for how common these words are in language in general, and in a corpus in particular. Some words (e.g., 'the', 'in', 'of') occur often in language in general and high frequency is an artefact of general language use. The simplest approach to dealing with this issue is the removal of these so-called **stop words** using lists of common terms in a language to retain only words thought more likely to contain meaning.

Of particular importance to environmental narratives are a family of methods collectively known as **named entity recognition (NER)**. These focus on identifying and classifying proper nouns such as the names of people, organisations and places. An important task in NER, and dealing with semantics more generally, is disambiguation. Lexical ambiguity refers to the phenomenon where a word has multiple possible meanings. Without additional context, we cannot resolve lexical ambiguity. For example, the word 'duck' can mean to crouch down or refer to a water dwelling bird. Given a POS tagger, and an associated sentence, these two meanings can be disambiguated since one sense of the word is a verb and the other a noun. A specific form of lexical ambiguity is referent ambiguity, where the same name is used to refer to multiple places (an extreme example is that there are more populated places named Springfield in the United States than there are US states). Assigning semantics to words is prone to errors which are often related to ambiguity.

Having identified named entities and dealt with referent ambiguity, they can be related to a unique instance (a specific person or place). By using external knowledge bases, such as place name gazetteers (Hill, 2009) containing information about these instances, we can add additional semantics to a text such as place types (e.g., village, forest, mountain) and other rich metadata (e.g., coordinates or bounding boxes) and link information to additional sources.

The quality of the tools used to perform these tasks varies greatly. For example, POS tagging in English is generally reliable. Dependency parsing within a sentence is highly effective, but linking entities across a narrative remains a difficult problem. NER is a vibrant research area, where much progress has been made, but often with a focus on particular classes of entity (such as organisations) and text genres (such as news reporting) and performance is often poor when methods are transferred to new genres or entity classes.

Named Entity Recognition

To demonstrate the potential and problems of an out-of-the-box solution for NER we processed two Geograph descriptions using the Python library spaCy[20]. The result of the first example is fully correct, 'Forestry Commission' is recognised as an organisation [ORG] and 'Balgownie' as a geopolitical entity or simply a location [GPE]. In the second example, we also see that many entities are labelled correctly, 'early January' is recognised as a date [DATE], 'North Norfolk District Council' as an organisation [ORG]. However, we also see that the locations were not recognised as such, 'Bacton Woods' and 'Witton Woods' are both wrongly labelled (as organisation [ORG] and person [PERSON] respectively). spaCy is

---

[20] https://spacy.io/usage/linguistic-features#named-entities

Corpus linguistics

Table 3.1 shows some basic properties of our corpus. Note how the number of tokens decreases based on what we treat as tokens. The lower value (not including punctuation) is more representative for most of the methods which will be applied here, since these are based on the BOW model described above, ignoring punctuation. Note also the mean (810) and median number of tokens per document (462). These numbers are especially interesting if we compare them to other corpora or, for example, if we compare different time periods within the same corpus. Information about the size of the corpus and its language should be always included in its description.

Table 3.2 illustrates, after normalising to lower case, token counts for the 20 most frequent words in our corpus. The words in the all tokens list convey no semantics with respect to the topic of the debates and include articles ('a', 'the'), prepositions (e.g., 'of', 'to', 'in'), conjunctions ('and', 'that'), verbs ('be', 'have'), and pronouns ('i', 'we'). Such words are typically included in stop word lists and removing these results in a revised frequency list. This list contains many words related to the general business of parliament. Some of these are obvious, e.g., 'government', 'people', 'minister', 'house'. Others require more knowledge of the language used in parliamentary debates, for example, members are referred to as "the right honourable member" or "my honourable friend" in speeches, and all of these words (or abbreviations thereof) appear as frequent tokens (e.g., 'hon', 'member', 'right', and 'friend'). This second list, we hypothesise, thus tells us something about the nature of parliamentary debates in general, but little or nothing about those discussing the Forestry Commission (apart from the obvious appearance of our search terms 'forestry' and 'commission' and, possibly, 'scotland', reflecting that country's much more forested nature).

Count	Total corpus	Mean per document	Median per document
All tokens	1450791	920	533
Tokens without punctuation	1277661	810	462
Sentences	48136	31	17

Table 3.1: Basic counts for corpus.

Rank	All tokens	Count	All tokens No stop words	Count
1	the	98,580	hon	5299
2	of	49,039	government	4062
3	to	41,450	commission	3570
4	and	32,112	one	3512
5	in	30,294	forestry	3405
6	that	28,290	land	2843
7	a	22,349	member	2436
8	is	21,433	people	2397
9	i	17,995	right	2303
10	for	15,006	many	2205
11	it	14,249	bill	2160
12	be	13,168	minister	2101
13	have	11,097	house	2038
14	not	10,155	new	1985
15	we	9927	friend	1971
16	are	9604	made	1871
17	on	9372	time	1845
18	which	9023	Scotland	1811
19	this	8667	great	1784
20	as	8279	years	1774

Table 3.2: 20 most frequent terms in corpus before and after filtering for stop words.

We can also look at term usage across a corpus by weighting the importance/relevance of a term for a document in comparison to its use in a corpus overall. One popular method is **term frequency-inverse document frequency (TF-IDF)**. TF-IDF is the product of the term frequency in a given document with the logarithm of the inverse fraction of the total number of documents in which the term occurs. It gives words common in all documents across the corpus lower scores than those which are common in a small sub-set of documents from the corpus. TF-IDF is a simple but very effective way of ranking the importance of terms in documents, where the corpus overall is used to calculate IDF, and in corpora, where another corpus (e.g., the British National Corpus described above) is used to estimate "normal" use of a term.

Having identified potentially interesting words, we can explore individual words qualitatively and quantitatively using a variety of methods. Perhaps the simplest is to use the idea of **concordance** to explore a word in context. Here,

of trees is to balance the features of the landscape. This is of interest to the town and country
e attitudes of farmers towards recreation and the landscape to be assessed. A characteristic of the project w
o is responsible for the liabilities. We want the landscape to be lifted and to see change, but because
e 10 Christmas bonus, provide a free tent? Is the landscape to change from wheat fields to caravan parks? Are
ere are four additional reasons. The first is the landscape value, and these two woods are situated in one
the 39 heritage coasts and in an area of great landscape value. The Government may despair of planners and
te more attention to increasing the beauty of the landscape. We propose to make no change in the structure
f Dean shares many of the characteristics of that landscape. We, too, have a verderers court and an area
for the future. Let us try to improve the landscape. What about the desecrations? Why do we not have
panese flowering cherries can replace a cherished landscape where English oaks have grown and matured over th
are causing a nuisance or distress.In a changing landscape, where hedgerows and other linear features that a
. It is a very poor policy in forestry or landscape, where you have to think of all time, to
the importance of humans in the history of the landscape, whether she was talking about the lido at Tootin
wish to say that this is a piece of landscape which has enjoyed public access for more than thr
most unpleasant things. They are a scar on the landscape which is slow to heal. They bring vehicles, somet
st topic concerns the countryside as a whole. The landscape which we enjoy today was substantially man-made,
they wanted and to enjoy the beauties of the landscape while at the same time observing the needs of
create a new Sherwood forest that would lift the landscape, yet the deal is stuck because there are question

Figure 3.2: Example concordances in the Hansard corpus.

the corpus as a whole is searched for all instances of a potentially interesting word, and these are then visualised in within the sentence or passage of text in which they occur.

Concordance

In Figure 3.2, example concordances with 'landscape' in our Hansard corpus are shown, highlighting the three words occurring after landscape. The sentences as a whole quickly reveal different ways in which landscape is discussed, for example with respect to its components, such as 'English oaks', 'hedgerows', 'woods', 'wheat fields'; change and value, such as 'change', 'desecration', 'lifted'; and access and recreation, such as 'public access', 'carvan parks'.

The notion of **co-occurrence** takes this one step further by looking at individual words which are found within a given distance of a search term or "node". For example, in the previous sentence the words 'a:2', 'given:1', and 'of:1' co-occur within a two word window of 'distance'. If we then remove stopwords we are left only with 'given'. Note that the order of steps is important here – if we first remove stopwords and then identify co-occurrences within a two-word window we find 'within:1', 'given:1', 'search:1', and 'term:1'. The influence of such seemingly minor choices can be very important, and reporting these choices is crucial if research is to be reproducible.

In a large corpus we can explore such co-occurrences in detail, and in particular look for meaningful combinations of words, termed **collocates**. We can find collocates in a corpus by looking for statistically significant co-occurrences. Statistical significance implies that the two words co-occur together more than we would expect by (random) chance given the overall frequencies of words in a corpus. We can use a wide range of measures to calculate statistical significance, all of which can be interpreted in slightly different ways. For example 'mutual information' favours exclusive and infrequent co-occurrence, while the 'T-Score' favours non-exclusive and frequent co-occurring terms (Brezina, 2018). Significant co-occurring terms can be ranked using simple frequency, or measures such as mutual information and T-scores. In analysing environmental narratives we are interested in not just statistically significant collocation, but also those which convey meaning in a specific context. Such phrases are recognisable in language, familiar to native speakers and seemingly logical substitutions sound clumsy or wrong. For example, though 'white mountains' and 'snowy mountains' contain similar information about the visual appearance of snow-covered distant mountains, the latter is a much more natural construction.

These approaches can be adapted by attaching more semantics to the corpus. Perhaps most simply, words can be merged to group those with the same meaning using normalisation, stemming and lemmatisation as described above. Normalisation approaches might include reducing all words to lowercase, creating canonical forms of words including diacritics, merging singular and plural forms of words and resolving different spellings to a single canonical form (e.g., 'colour' vs. 'color'). All of these methods can be further refined by also including information about parts of speech through **part-of-speech (POS) tagging**. A POS tagger assigns each word in a document a tag, such as verb, noun, adjective, preposition and so on. A related family of tools, dependency parsers analyse the grammatical structure of sentences and transform them to so-called dependency trees.

Dependency parsing

We used the Python library spaCy[18] to process the following fictive sentence: 'Beautiful, peaceful landscape in Bacton Woods'. The dependency tree shown in Figure 3.3 includes dependency labels (amod - adjectival modifier, prep - prepositional modifier, pobj - object of preposition, and compound) and part of speech tages for each words (ADJ - adjective, NOUN, ADP - adposition, PROPN - proper noun).

The compound place name 'Bacton Woods' is correctly recognised as such and the words it consists of are correctly labelled as proper nouns. Both 'beautiful landscape' and 'peaceful landscape' are identified as adjectival modifiers. Such compounds can be useful as features for machine learning as will be described in more detail in Section 3.4.2.

---

[18] https://spacy.io/api/annotation#dependency-parsing

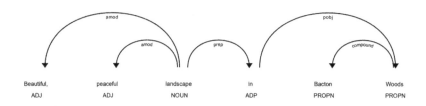

Figure 3.3: Sample dependency tree.

Comparison with a general corpus of English

One approach to finding more semantically interesting terms in the frequency lists discussed above is to compare frequent adjectives in our debates to frequencies in a general corpus of English (the British National Corpus [BNC]), which contains both written and spoken English). In Figure 3.4 we plot the ranks of the 20 most

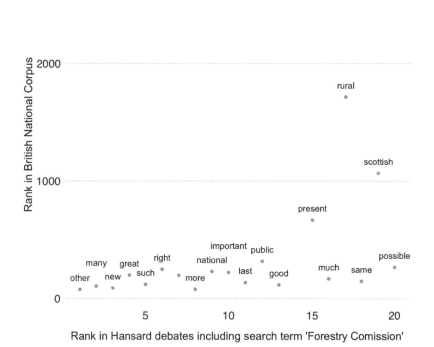

Figure 3.4: Adjectives ranked in BNC and the Hansard corpus.

frequent adjectives in our Hansard corpus against their corresponding ranks in the BNC. Many of these adjectives are relatively common in the BNC, with 16 of the 20 in the most 500 frequent words found in the BNC. Three however, appear to be much rarer in general text, 'scottish', 'rural', and 'agriculture', all of which have ranks higher than 1000. These frequencies suggest that agriculture and rural landscapes are often discussed in relation to the Forestry Commission, and that Scotland often appears to be referred to in this context.

These results are obvious, but they illustrate how we can pre-process our corpus to gain first insights into important topics. The choices we make along the way (e.g., in normalising text, using POS taggers and stemming word to their roots) all influence results in both predictable ways (e.g., stemming "year" and "years" will aggregate all counts referring to time in this way) and less obvious ways. For instance, roughly 50% of the instances of forestry were classified as nouns and 50% as adjectives.

Co-occurrence and TF-IDF

Another way to finding semantically interesting terms is to use TF-IDF. However, first, since our initial hypothesis was that discussions of the Forestry Commission would focus on negative impacts on landscape, we decided to explore co-occurrence with landscape. As explained above, we had to choose some parameters for our search window in the corpus. We decided to retrieve all instances where a word co-occurred more than three times with landscape, in a search window spanning three words before and after our node (landscape). We ranked co-occurring terms by both raw frequency and mutual information (recall this measure favours exclusive, infrequent co-occurrence) (Table 3.3).

The raw frequencies reveal that landscape is often referred to in terms of beauty, and by looking at ranks based on mutual information we see that this relationship is indeed specific to landscape and is captured in related words such as 'beauties' and 'beautiful'. Furthermore, the importance of change and preservation is captured by terms such as 'change', 'lift', 'enhance', and 'unique'. These results suggest that landscape does indeed appear to be discussed in the context of impacts.

TF-IDF can reveal the relative importance of these words and provide some hints as to context. For example, the very first mention of 'landscape' in our texts in 1925 (1925-03-30a.992.4.txt)

Co-occurring term ranked by frequency (count)	Co-occurring term ranked by mutual information (count)
beauty (13)	lift (3)
change (9)	coverage (4)
rural (7)	beauties (3)
forestry (6)	accustomed (3)
countryside (6)	enhance (4)
features (6)	beauty (13)
areas (5)	features (6)
farmers (5)	unique (3)
coverage (4)	beautiful (3)
hon (4)	wildlife (3)

Table 3.3: Ten most frequent terms co-occurring with landscape ranked by frequency and mutual information.

seems to be in the context of roads with the following terms highest ranked according to TF-IDF: ['roads', 'road', 'fund', 'tolls', 'unclassified', 'ought', 'gbp', 'blind', 'corners', '42000000']. A later text, reporting a debate in 1977 (1977-05-05a.705.4.txt), seems to focus on recreation, since its highest ranked TF-IDF terms are: ['recreation', 'sport', 'recreational', 'facilities', 'education', 'sports', 'schools', 'regional', 'enthusiasm', 'fringe'].

However, to understand the whole picture we need to move from a bag of words representation, or "macroanalysis", and start to explore in more detail ways in which landscape has been discussed in parliamentary debates over time through 'microreading'.

Reading of individual texts, often termed in the context of computational analysis **microreading**, is an obvious, and important, but sometimes neglected task (Jockers, 2013)[19]. It involves, as discussed in the introduction, reading and interpreting individual passages or texts that have been identified as potentially of interest computationally, and is closely related to the notion of hermeneutic analysis introduced earlier. In microreading, we use our knowledge of the corpus and associated context to interpret and refine our understanding of text.

---

[19] Macroanalysis and microreading are also often termed distant and close reading (Moretti, 2013). We prefer the macroanalysis/ microreading distinction here, since it emphasises that computational methods often do not involve any reading in a traditional sense, but rather a macroanalysis of, for example word frequencies.

Microreading

If we zoom into the texts we identified above, then we quickly find that the speaker in 1925, Lieut-Colonel Wilfrid Ashley of the Conservative Party, was, as suggested in the TF-IDF analysis, concerned with the link between roads, the landscape, and trees. Only by reading the text though do we see how his speech mirrors the original text analysed by Graham Fairclough, discussing roads in terms of both progress and aesthetics, and proposing trees as a way of mitigating their ugliness.

'The origin of this Bill is rather quaint. The year before last, when I had the honour of being Parliamentary Secretary to the Ministry of Transport, it was brought home to me very forcibly that these new roads which had been constructed, and were being constructed in the vicinity of the Metropolis, however excellent they might be from the transportation point of view and however useful from the national aspect are extremely ugly. I think the House will agree with me that a great wide stretch of road surface, in most parts bounded by concrete posts bound together by iron wires, is not a very graceful or grateful addition to the landscape. So I went into the matter rather fully, and came to the conclusion that, if proper trees be planted alongside some of these great roads, it would, at any rate, in a few years take off the bareness of the aspect and replace many trees which had had to be cut down when these new roads were made.'

The second text, a speech by Mr Kenneth Marks (Labour Party) also demonstrates nicely that TF-IDF correctly identified the importance of recreation. However, contrary to our original expectations, the Forestry Commission is here being discussed in a positive sense with respect to its impact on landscape.

'The commission has been helping with the reclamation of derelict land in river valleys, tree planting, general landscape improvement works and provision for informal recreation facilities, such as picnic areas, footpath and bridleway systems, and information facilities. The commission has also engaged the Civic Trust for the North-West to carry out an experimental project to promote recreational use, to encourage conservation and to stimulate local interest in the Tame Valley.'

Reading both of these texts also reveals something of the formal and rather complex nature of parliamentary language, which has also changed over time, a further issue for consideration in their analysis.

trained on the corpus OneNotes Release 5.0[21], a collection of news, weblogs and transcribed telephone conversations. This example shows that this out of the box solution does not work perfectly on some of our texts, and that either use of additional rules or retraining the algorithm on annotated texts would be necessary.

Example 1:
Thinning out at Forestry Commission **ORG** mixed woodland at Balgownie **GPE** [22].

Example 2:
... are the prevalent woodland colours in early January **DATE** . Bacton Woods **PERSON** , also known as Witton Woods **ORG** , covers 113 **CARDINAL** hectares; the woodland is owned by the Forestry Commission **ORG** and partly managed by North Norfolk District Council **ORG** , who together form the Bacton Woods Countryside Partnership Project **ORG** [23].

Irrespective of the tool being used, manual annotation remains the gold standard for adding semantics to a corpus. Annotation is a time consuming but very valuable way of analysing a corpus. It can be used directly as an analytical tool, to validate results produced computationally, or to create training data used in machine learning approaches. Regardless of the application, annotation requires a set of rules defining categories for annotation and the rules used to identify them, some form of replication by multiple annotators (typically reported in the form of inter-annotator agreement after annotation of the same or overlapping corpora by independent annotators using the same rules), and a way of storing annotated texts for future use. Since annotation is a very common activity, many community standards already exist, such as the Text Encoding Initiative (TEI) guidelines[24]. Annotation tools aim to make the annotation process simpler, and are increasingly moving to online tools such as Inception (Klie et al., 2018) and Recogito (Simon et al., 2017), developed with a focus on annotation in the Spatial Humantities and compatible with common formats including TEI[25]. Annotation is closely related to "microreading" since both involve a detailed reading of the text, with the main difference relating

[21] https://catalog.ldc.upenn.edu/LDC2013T19
[22] Paul McIlroy, https://www.geograph.org.uk/photo/285266
[23] Evelyn Simak, https://www.geograph.org.uk/photo/650293
[24] https://tei-c.org/release/doc/tei-p5-doc/en/html/index.html
[25] https://recogito.pelagios.org/

to purpose – microreading is often concerned with a qualitative interpretation, while annotation results are used as training, test and evaluation data in quantitative work. In practice, many projects develop bespoke annotation schemes specific to the task at hand, and Pustejovsky and Stubbs (2013) give a useful overview of a potential pipeline which they call the model-annotate-model-annotate cycle, emphasising the importance of iteratively modelling (i.e., specifying the concepts that should be annotated) and actually annotating data.

Text Encoding Initiative

The annotation can be done on different hierarchical levels, for example, we can annotate 'early January' simply as <date> or we can add elements 'notBefore="–01-01" notAfter="–01-10"' indicating when the date (or in this case time period) actually is. Similarly, the tag <placeName> can contain more detailed information, for example, it can be divided on <settlement> and <region> (Listing 3.2).

Listing 3.2: TEI example of a Geograph description contributed by Evelyn Simak.

```
<TEI xmlns="http://www.tei-c.org/ns/1.0">
<teiHeader>
<!---...-->
</teiHeader>
<text>
<body>
<l>
... are the prevalent woodland colours in
<date notBefore="--01-01" notAfter="--01-10">
early January</date>.
<placeName>Bacton Woods</placeName>,
also known as <placeName>Witton Woods</placeName>,
covers 113 hectares; the woodland is owned by
<orgName>the Forestry Commission</orgName> and
partly managed by <orgName>North Norfolk District
Council</orgName>, who together form the
<orgName>Bacton Woods Countryside Partnership
Project</orgName>.
</l>
</body>
</text>
</TEI>
```

3.4.2   Classification

Often we are interested in analysing texts grouped by a common topic, gender of the writer, time period and so on. To do so, we can either perform **unsupervised classification** by grouping texts based on their statistical similarities and adding the labels to the emerging classes, or **supervised classification**, where classes are defined in advance.

For **supervised classification**, training data has to be either already available or created prior to the classification, often through the process of annotation, sketched above. For classification tasks, a typical annotation workflow includes the following steps:

- Identification of desired classes.
- Creation of the set of clear rules allowing independent annotators to annotate texts consistently. Commonly, a small random sample of the data is selected to refine the rules and give examples.
- Independent annotation and calculation of inter-annotator agreement. For this task around 10% of the randomly selected texts are suitable. Inter-annotator agreement is calculated using a statistical measure, typically Cohen's or Fleiss' Kappa, depending on the number of annotators (Landis and Koch, 1977; Pustejovsky and Stubbs, 2012).
- If inter-annotator agreement is acceptable for type of texts (e.g., lower Kappa is acceptable for complex historical texts) one annotator can proceed with the rest of the annotation. Otherwise, the rules should be refined, another random 10% selected and annotated until inter-annotator agreement reaches the desired value.

Annotation

Since our initial hypothesis was that the Forestry Commission's actions are perceived as leading to negative impact on landscapes, we set out to classify all the Hansard texts containing the word 'landscape' into three classes: 'negative impact', 'positive impact' and 'neutral'. We expected the Hansard texts to contain a limited number of clearly opinionated texts towards the Forestry Commission. Therefore, we added another collection of texts – Geograph[26] – to enrich our corpus with texts more likely to contain positive or negative sentiments.

---

[26] https://www.geograph.org.uk/

To accomplish this task we, firstly, created a set of the following rules:

- 'Negative impact' included descriptions of current and former negative consequences of the Forestry Commission strategies, infrastructure or negative references to Forestry Commission buying of land practices. Examples: *The horrible trees on the left are privately managed and the horrible trees on the right are in a Forestry Commission holding*[27]*.; This is the sort of planting which got Forestry Commission woodland such a bad name*[28].
- 'Positive impact' included texts describing positive influence on landscape, such as actions towards revival of native wood, positive effect on biodiversity and creation of infrastructure for recreation. Examples: *The Forestry Commission are encouraging the regrowth of natural woodland species in the Knapdale Forest*[29]*.; This area of heathland and bog would be inaccessible to walkers without footbridges like this one, constructed by Forestry Commission engineers*[30].
- 'Neutral' descriptions include factual statements or describe effects on landscape without positive/negative judgement. Examples: *A Forestry Commission house in Penninghame Forest*[31]*.; The forest had been replaced by spruce plantations here by the Forestry Commission. Policies have changed and this area is likely to revert to oak in the future, now that the spruce has been removed*[32].

Two annotators then annotated 15 Hansard texts according to the rules described above. Eleven of 15 descriptions (73%) were identically annotated by both annotators; however, Cohen's Kappa of 0.58 is moderate (Table 3.4), therefore, a further random 15 texts were selected, for which the annotators reached the substantial agreement of 0.78.

Cohen's Kappa is calculated based on a confusion matrix (Table 3.5), where the results of one annotator are

---

[27] *Richard Webb, https://www.geograph.org.uk/photo/166532*
[28] *Barbara Cook, https://www.geograph.org.uk/photo/104209*
[29] *Patrick Mackie, https://www.geograph.org.uk/photo/245251*
[30] *Jim Champion, https://www.geograph.org.uk/photo/92418*
[31] *Oliver Dixon, https://www.geograph.org.uk/photo/172754*
[32] *Richard Webb, https://www.geograph.org.uk/photo/187987*

Kappa statistic	Strength of agreement
< 0.00	Poor
0.00–0.20	Slight
0.21–0.40	Fair
0.41–0.60	Moderate
0.61–0.80	Substantial
0.81–1.00	Almost perfect

Table 3.4: Relation between Kappa statistics and strength of agreement as proposed by Landis and Koch (1977).

	Negative	Neutral	Positive	Row totals
Negative	3	0	0	3
Neutral	2	6	2	10
Positive	0	0	2	2
Column totals	5	6	4	15

Table 3.5: Confusion matrix of the first annotation of the Forestry Commission text into three classes: negative, neutral, positive.

written horizontally, and the other vertically. Then, it is calculated according to the following formula:

$$k = \frac{\sum_a - \sum_{ef}}{n - \sum_{ef}},$$

where $\sum_a$ is sum of the agreements (the diagonal), n – total number of texts, and $ef = \frac{row\_total * column\_total}{overall\_total}$ – expected frequency per class.

Sum of the agreements:

$$\sum_a = 11$$

Total number of texts:

$$n = 15$$

Expected frequencies:

$$ef_{negative} = \frac{3 * 5}{15}, ef_{neutral} = \frac{10 * 6}{15}, ef_{positive} = \frac{2 * 4}{15}$$

$$\sum_{ef} = 5.53$$

Cohen's Kappa:

$$k = \frac{\Sigma_a - \Sigma_{ef}}{n - \Sigma_{ef}} = \frac{11 - 5.53}{15 - 5.53} = 0.58$$

Through the process of manual annotation the following distribution of texts according to classes emerged:

- negative 9
- neutral 24
- positive 13

Following the same rules, a single annotator annotated all Geograph texts containing 'Forestry Commission', which after filtering for identical descriptions contributed by the same author resulted in 3014 texts. The majority of the Geograph texts were also neutral:

- negative 105
- neutral 2687
- positive 322

However, it is important to note that many of the texts are not neutral in the traditional sentiment analysis sense. For example, the following description clearly shows negative sentiments towards Ordnance Survey mapping decisions, but does not provide any information about acceptance of the Forestry Commission actions: *A purple mess cluttering up the map states that this is Forestry Commission land*[33].

Having annotated data, for example into binary (e.g., positive, negative), nominal (e.g., forest, meadow, urban, lake) or ordinal (e.g., sentiment ranging from very negative through neutral to positive) classes, then it is possible to fit statistical models to text features or train machine learning models using text features. The most common representation of a text is through the so-called feature vectors (see Section 3.4.1). The simplest feature we can use in text processing is a vector containing zeros and ones representing absence and presence of the n most frequent unigrams in the whole corpus.

For example, if the five most frequent unigrams in a corpus, after stop word removal are 'timber', 'recreation', 'tree', 'beach' and 'sea', then a text mentioning only 'timber' will be represented as the vector $[1, 0, 0, 0, 0]$, and texts mentioning only 'recreation' and 'tree' will be represented as $[0, 1, 1, 0, 0]$. Other features could include (typically normalised) frequency of unigrams, frequency of other n-grams, number of words belonging to the same POS (e.g., adjectives),

---

[33] https://www.geograph.org.uk/photo/1826512

or frequency of defined syntax dependencies (e.g., adjectival modifiers). Thus, feature vector representation of the following sentence: *Beautiful, peaceful landscape in Bacton Woods* based on frequency of nouns, adjectives and total number of words is [*n_nouns, n_adjectives, n_words*] or [3, 2, 6]. Other common feature types include presence or absence of words from lists of relevant terms (e.g., sentiment lexicons containing terms commonly associated with positive or negative sentiment) or more complex compound features, for example capturing the similarity of texts.

Having encoded texts into features, we can apply a variety of statistical and machine learning methods to predict how other documents should be classified, including general linear models, random forests, naïve Bayes, support vector machines and neural networks. In practice, some classifiers work better than others for text data—and some work better on smaller datasets or can be trained more quickly, while others are more effective on very large datasets. Naïve Bayes is a relatively simple probabilistic model that assumes that the words used in a document are statistically independent of one another. This assumption of independence means that naïve Bayes is prone to error but it also can discover words that are important indicators of a category even in quite small data sets. It is also very fast to train.

Overfitting is an important problem in machine learning, where good performance is possible on a training data because the model slavishly fits to individual data points, and thus does not generalise well when presented with an unseen set of feature values. An example of a classifier not prone to overfitting is the random forest classifier, since it creates random sub-sets of the features and builds smaller trees using these sub-sets. In contrast, more sophisticated classifiers using, for example, neural networks, are highly effective for large corpora, but they are prone to overfitting the training data when working with smaller data sets, and they require extensive computational resources to train.

Evaluation of such models can be carried out in a number of ways. Very common are the calculation of **precision**, **recall** and **F1** scores. Precision is the proportion of correctly classified texts. For example, in a corpus of 20 texts, if 10 texts were classified as positive but only eight were annotated as positive, then the precision would be 8/10 (0,8). Recall is a measure for the completeness of a result, and is the proportion of texts belonging to a class which we return. Thus, if a total of 16 texts were annotated as positive in our example, then the recall would be 8/16 (0.5). The F1-score is the harmonic mean of precision and recall.[34] In our case, F1 would therefore be 0.62. F1 is particularly useful in comparing performance of classifiers with different feature sets, but less illuminating in isolation. Depending on the task at hand we may choose to optimise for precision, aiming to have a classifier which makes as few mistakes as possible, or recall, returning as many relevant examples as possible.

---

[34] Calculated as $F1 = \frac{2 \cdot precision \cdot recall}{precision + recall}$.

In closing this section, it is worth remembering one final point. In recent years awareness has greatly increased of the environmental impacts of all aspects of human behaviour, and computational methods are no different. Considering the potential environmental impacts of, for example, training machine learning approaches to classification is an important consideration (Bender et al., 2021).

Classification

The Countryside Act 1968 extended the powers of the National Parks Commission, renaming it the Countryside Commission, and extending its conservation and recreational remits. It also gave the Forestry Commission the explicit power to enhance access to forests for enjoyment and recreation. Since all of our texts are attributed with a date, and 1968 falls almost exactly mid-way through the time period captured in our Hansard corpus, we hypothesised that changes in the language used with respect to debates might allow us to classify texts according to their data of publication. Since data of publication was given as metadata, we could use this date directly to group debates in two classes 'before' and 'after' 1968.

To do so we randomly divided all our texts on two halves of training and test data. Training data consisted of 416 texts before and 373 after the year 1968, while test data contained 421 texts before and 367 after 1968. We then trained a random forest[35] on our training data based on a range of features to classify texts and evaluated model performance on our test data set.

Using only the 300 most frequent unigrams, our model has an F1 of 0.808, already a relatively good performance, suggesting that language does change between these dates. After filtering out very short descriptions, containing less than 20 unigrams, F1 increased very slightly to 0.810. Using bigrams (e.g., 'climate change') and length of descriptions as features did not improve the prediction ability of the model. We had hypothesised that these features might be effective in capturing, on the one hand, new issues such as climate change, and on the other more or less controversial topics (as opposed to simple descriptions). We suspect that the total number of texts was too small for these to improve model performance in this case.

---

[35] https://scikit-learn.org/stable/modules/generated/sklearn.ensemble.
RandomForestClassifier.html

An additional useful property of random forests is that a measure of the relative importance of each feature on the prediction is returned. Since features here are simply vectors of unigrams, we can explore which unigrams are most likely to allow us to classify texts temporally. The 25 most useful unigrams were: *environment, asked, constituency, present, ask, quite, change, end, like, countryside, management, work, friend, regard, men, committee, timber, acres, kind, important, public, secretary, government, agriculture, land.* Some of these (e.g., 'environment', 'timber', 'change', 'countryside', 'public') may reflect changes in the topics being discussed with respect to forests, and potentially a move towards recreation and conservation and away from timber production.

About 70 descriptions in each class were classified wrongly. Many of these texts belong to the years around 1968. It is clear that themes of debates do not change sharply in 1968. However, the greatest number of wrongly classified texts (eight) was in 1978. A microreading analysis of these texts shows that many wrongly classified texts in 1978 mention the Forestry Commission in passing as one of examples of organisations as in the text below (1978-03-21a.1462.1.txt):

'Those powers at present are exercised not by Ministers of the Crown but by bodies such as the Forestry Commission and the Housing Corporation as specified in Schedule 7'.

Other texts refer to issues returned to throughout the history of the Forestry Commission as in the example below (1978-07-20a.904.2.txt):

'The total area of existing Forestry Commission forest in Wales is almost one-third of that in Scotland, but if we make allowance for Scotland's greater size we find that the proportion of land afforested is about the same'.

Using the simple features we selected for illustration here, we could not accurately predict the time of writing of such texts. In a typical iterative process, we might add additional features based on this microreading (e.g., presence of names of government organisations) to our feature vectors. In doing so however, it is important to retain 'unseen' data on which we test a final model. In this example we do not demonstrate an exhaustive list of such features, nor do we go beyond simple unigrams to more advanced features such as TF-IDF scores for terms, since our aim was to illustrate that a classifier can distinguish between two periods using simple features.

Regression

Since the metadata associated with debates is given on an interval scale as a year, we can also treat this problem as one of regression, and attempt to predict the year of a debate. Random forests can be used for both classification and regression[36], and making this change is straightforward computationally if the data provided furnish the necessary classes.

Using the same features as for our binary classifier (300 most frequent unigrams and filtering short texts with less than 20 tokens), we could train a random forest regression model with an r^2 of 0.413. This implies that about 40% of the variation in date attributed to a debate can be predicted by the choice of words alone, independent of their detailed context. Delving more deeply into the results, we note that only 168 descriptions (ca. 21%) are assigned a date with an error of more than 20 years. Once again, microreading, is an important way of exploring our corpus. For example, two contributions, from 2019 (2019-03-28b.545.11) and 1942 (1942-07-28a.330.7) were predicted with large errors of 1932 and 1990, respectively.

'Bishop Wood is being used for shooting—land leased by the Church Commissioners to the Forestry Commission. Blood sports in exchange for blood money for the Church of England. What steps have the Church Commissioners taken to ban blood sports across their estate?' (2019-03-28b.545.11)

'Sixty-three per cent. of the officials employed in Forestry Commission plantations in Wales are Welsh. Consequently there is not a preponderating number of English. Welsh officials are also employed in England, such interchanges being both desirable and necessary in the interests of the Forest Service as a whole.' (1942-07-28a.330.7)

Both of these texts are short, and without additional contextual information we suggest difficult or impossible for a human to date in a meaningful way. With additional information, the former text, discussing as it does 'blood sports' and related to the controversial ban and discussion around hunting in the UK in the early 21st century can easily be dated, but once again the features used in our model are not capable of identifying such changes. Rather we suggest, that such outliers can provide informative ways of zooming in and out from our corpus and identifying emerging themes of potential interest.

---

[36] https://scikit-learn.org/stable/modules/generated/sklearn.ensemble.
RandomForestRegressor.html

In contrast to supervised classification methods which require annotated training data, **unsupervised methods** require no training data. Such methods are often used to explore corpora, and can provide powerful and straightforward ways of identifying common threads of discourse within a corpus. Perhaps the most well known such family of methods is topic modeling (Blei, 2012). One very commonly used form of topic modeling is **latent Dirichlet allocation (LDA)** (Blei, Ng, and Jordan, 2003). The basic notion, if not the mathematics underlying the approach, is relatively straightforward. Imagine a corpus of documents derived from a newspaper, where each article is stored as a document. Different newspapers publish different genres of articles, ranging, for example, from sports reporting through editorials and travel reporting to celebrity gossip, political reporting, local news and foreign affairs. A given article might though combine aspects of these genres, for example, a story reporting on Brexit negotiations combines both political reporting and foreign affairs. LDA, given the raw text of articles, attempts to do two things:

1. Identify a set of n topics which best differentiate individual documents based on the bag of words model;
2. Assign to every term in the probability that it belongs to a given topic.

The set of topics generated are based on the co-occurrence of terms in documents, and are often claimed to be easily interpretable (Chang et al., 2009). A new, unseen, document can then be associated with one or more topics, based on the terms making it up and their probability of belonging to individual topics. Topic modeling can therefore be used in three distinctive ways. Firstly, topic modeling can be used to explore a corpus. By generating a set of topics, examining the terms making up a topic and assigning labels to topics it is possible to in principle identify different forms of discourse. Importantly, the explorative process is sensitive to a range of input parameters, including crucially the number of topics and to the ways in which the corpus is pre-processed. Secondly, topic modelling can be used predictively, analogously to the supervised methods described above. For example, performing topic modelling and identifying three classes of readers' letters: those supportive of a government, those critical of a government and those discussing other matters. Given a new letter, we could then identify to which, if any, topic it best belonged. Finally, it is possible to use topic modelling to find semantically similar documents. For example, given a document that contains a specific mixture of topics, we can find other documents that share that particular mixture, or do so while also adding in another additional topic.

Topic modelling relies on the distributional hypothesis – neatly summed up by the linguist Firth in 1957 as 'You shall know a word by the company it keeps!'. The critical reader will, we hope, note that this also implies some dangers inherent in topic modelling. Topic modelling relies entirely on a bag of words model, and as we have seen the language used in different domains can vary considerably. Thus, language might not vary only according to the nature of a debate, but according to the domain or genre of writing, or indeed according to the backgrounds of the authors. Given, for example, a corpus of nature descriptions written by school children and adults, we might expect the age of the authors to be more decisive in determining topics than the content of the descriptions themselves.

Topic Modelling

In the classification step we annotated Geograph descriptions into three classes: negative, neutral and positive. The decision about the nature of the classes was made beforehand according to our hypothesis. However, these descriptions cover a variety of other topics and can be classified in many different ways. To explore these possibilities, we used a Python implementation of LDA[37]. One of the important decisions in topic modelling is the number of classes. There are methods to approach this problem quantitatively, but we simply experimented with 20 topics, and below are three examples.

Topic 1: Cycling/walking
Most probable words: *park, walkers, road, entrance, bit, car, narrow, route, signs, cycle, woodland, cycling, forest, heads, popular, lodge, farm, walking, land, village*

Example descriptions:
Looking east towards the Royal Oak pub, the "centrepiece" of Fritham *village*. The place is always busy with visitors on weekends and during the holidays. There are several Forestry Commission *car parks* which provide convenient access to the surrounding *Forest* (on foot, bike or horse)[38].

The tracks in the Forestry Commission *land* in the New *Forest* are very *popular* with *cyclists* and *walkers*[39].

Topic 2: Second World War airfields
Most probable words: *operated, road, mor, monadh, woodland, forest, wind, car, park, part, WWII, hill, following, used, airfield, loch, warning, timber, track, area*

---

[37] https://radimrehurek.com/gensim/
[38] Jim Champion, https://www.geograph.org.uk/photo/69306
[39] Nigel Mykura, https://www.geograph.org.uk/photo/6361160

Example descriptions:
Used by the Forestry Commission for storing *timber* & compost; a concrete approach *road* probably shows its origins as *WWII airfield* use[40].

The Forestry Commission *car park* at Janesmoor Pond is on the site of a Second World War *airfield* – odd bits of brick and concrete remain here and there. The gravel surface of the *car park* overlies the former service roads on the *airfield*. It is a very spacious *car park*, by New *Forest* standards, capable of accommodating large vehicles with horse trailers[41].

Topic 3: Access
Most probable words: *wood, road, country, access, park, also, route, woodland, public, forest, area, part, carved, track, footpath, car, conifer, mostly, plantation, accessible*

Example descriptions:
Although Forestry Commission, this *wood* is not mapped as *public access*. Presumably only leasehold – a confusing distinction to the *public*. However, a short distance along this *track* it is joined by a public *footpath* which starts at a different point on Fisher Lane[42].

Part of the Callan's Lane *Wood* Forestry Commission *public access* scheme, this *wood* forms the central *part* of the mixed *woodland*. This grass *track* leads to open farmland, with mixed broadleaved trees on the left and *conifers* on the right[43].

3.5   Where to Next?

Our aim in this chapter was to introduce a methodological tool box for undertaking the computational analysis of environmental narratives. This tool box contains not only concrete tools, such as those for part of speech tagging or NER, but also requires that we think about the questions we (can or should) ask of texts, narrative forms used in text, ways of sourcing or building corpora and often forgotten issues of copyright and ethics with respect to our sources.

As an example we set out to explore ways in which the Forestry Commission was discussed in the UK over the last 100 years in two contrasting corpora: speeches from the House of Commons and a collection of crowdsourced image descriptions. We hypothesised that the perceived negative impacts of the

---

[40] Mike Faherty, https://www.geograph.org.uk/photo/3884578
[41] Jim Champion, https://www.geograph.org.uk/photo/62968
[42] Robin Webster, https://www.geograph.org.uk/photo/335443
[43] Kate Jewell, https://www.geograph.org.uk/photo/405217

Forestry Commission on landscape would be visible in parliamentary debates, and that the visual impacts of forestry would be emphasised in the image descriptions. The methods we used demonstrate that landscape, and the impact of trees and forestry on it, were commonly discussed in our corpus. However, extensively annotating texts revealed that in practice many were neutral, and more texts in both collections were positive than negative. Automatic analysis of the Hansard corpus showed that language use does allow us to differentiate between texts, and indicated that the nature of issues discussed has changed over time, with something of a move towards recreation. Interestingly, the importance of recreation and access also emerged in the topic modelling of the Geograph collection, where contrary to our expectations not only what could be seen was discussed, but also the influence of forestry on access (thus implying that what the photographers do in a location was important) as well as historical land use (the Second World War airfields, perhaps reflecting the interests of the contributors).

Perhaps the most obvious result of our exploration is the importance of context and a constant interplay between source texts and computational analysis in our interpretation. This observation closes the circle of this chapter, returning us to the importance of the hermeneutic circle and emphasising the importance not of the tools we use to read, but rather the ways in which we combine these tools to gain knowledge.

3.6   Suggested Readings

Literary theory and spatial language

Bleicher, Josef (2017). *Contemporary Hermeneutics: Hermeneutics as Method, Philosophy and Critique*. Vol. 2. Routledge.

This book by Bleicher is an excellent introduction to the theory of hermeneutics for literary criticism.

Lakoff, George, and Mark Johnson (2008). *Metaphors we live by*. University of Chicago Press.

In this influential book Lakoff and Johnson show the deep role of metaphors, including spatial metaphors, in how people communicate through language.

Lotman, Yuri M (1990). *Universe of the Mind: A Semiotic Theory of Culture*. London: IB Taurus.

Lotman provides a view of narrative analysis from a cultural semiotics perspective with his approach to the topic of literary space being of particular relevance to environmental narratives.

Digital literary analysis

Moretti, Franco. Distant reading. Verso Books, 2013.

Jockers, Matthew L (2013). *Macroanalysis: Digital Methods and Literary History*. University of Illinois Press.

Both of these books explore how statistical and computational methods can be used to perform literary analysis on narrative texts. They make the case for 'reading' narratives in aggregate to understand the sociology of literature in new ways.

Corpus linguistics and construction

McEnery, Tony, and Andrew Hardie (2011). *Corpus linguistics: Method, Theory and Practice*. Cambridge University Press.

Kennedy, Graeme (2014). *An Introduction to Corpus Linguistics*. Routledge.

These are comprehensive textbooks that describe the use of corpus data to study language. Topics include both the construction of corpora and issues of ethics as well as methods of analysis.

Natural language processing

Manning, Christopher, and Hinrich Schutze (1999). *Foundations of Statistical Natural Language Processing*. MIT Press.

Although the first edition of this book came out over 20 years ago, it remains an influential text and introduces the key statistical foundations for modern natural language processing.

Bird, Steven, Ewan Klein, and Edward Loper (2009). *Natural Language Processing with Python: Analyzing Text with the Natural Language Toolkit*. O'Reilly Media, Inc.

This book provides an excellent introduction to practical tools for doing natural language processing using the Python programming language. All the computational methods described earlier in this chapter are represented with examples.

Blei, David M (2012). Probabilistic topic models. *Communications of the ACM* 55, no. 4: 77-84.

In this summary article, Blei introduces the family of probabilistic topic models with clear examples of their use.

References

Alex, Beatrice, Claire Grover, Jon Oberlander, Tara Thomson, Miranda Anderson, James Loxley, Uta Hinrichs, and Ke Zhou (2016). "Palimpsest: Improving assisted curation of loco-specific literature". In: *Digital Scholarship in the Humanities* 32.November 2016. ISSN: 2055-7671. DOI: https://doi.org/10.1093/llc/fqw050.

Augenstein, Isabelle, Leon Derczynski, and Kalina Bontcheva (2017). "Generalisation in named entity recognition: A quantitative analysis". In: *Computer Speech and Language* 44, pp. 61–83. ISSN: 10958363. DOI: https://doi.org/10.1016/j.csl.2017.01.012.

Bender, Emily M., Timnit Gebru, Angelina McMillan-Major, and Shmargaret Shmitchell (2021). "On the dangers of stochastic parrots: Can language models be too big?" In: FAccT '21. Virtual Event, Canada: Association for Computing Machinery, pp. 610–623. ISBN: 9781450383097. DOI: 10.1145/3442188.3445922.

Blei, David M (2012). "Probabilistic topic models". In: *Communications of the ACM* 55.4, pp. 77–84. DOI: 10.1145/2133806.2133826.

Blei, David M, Andrew Y Ng, and Michael I Jordan (2003). "Latent dirichlet allocation". In: *Journal of Machine Learning Research* 3.Jan, pp. 993–1022. DOI: 10.5555/944919.944937.

Boell, Sebastian K and Dubravka Cecez-Kecmanovic (2010). "Literature reviews and the hermeneutic circle". In: *Australian Academic & Research Libraries* 41.2, pp. 129–144. DOI: 10.1080/00048623.2010.10721450.

Brezina, Vaclav (2018). *Statistics in corpus linguistics: A practical guide.* Cambridge: Cambridge University Press. DOI: 10.1017/9781316410899.

Butler, James O, Christopher E Donaldson, Joanna E Taylor, and Ian N Gregory (2017). "Alts, Abbreviations, and AKAs: Historical onomastic variation and automated named entity recognition". In: *Journal of Maps and Geography Libraries* 13.1, pp. 58–81. DOI: 10.1080/15420353.2017.1307304.

Chang, Jonathan, Sean Gerrish, Chong Wang, Jordan L Boyd-Graber, and David M Blei (2009). "Reading tea leaves: How humans interpret topic models". In: *Advances in neural information processing systems*, pp. 288–296.

Davies, Clare (2013). "Reading geography between the lines: Extracting local place knowledge from text". In: *Lecture Notes in Computer Science (including subseries Lecture Notes in Artificial Intelligence and Lecture Notes in Bioinformatics)* 8116 LNCS, pp. 320–337. ISSN: 03029743. DOI: 10.1007/978-3-319-01790-7-18.

Egorova, Ekaterina, Thora Tenbrink, and Ross S. Purves (2018). "Fictive motion in the context of mountaineering". In: *Spatial Cognition and Computation* 18.4, pp. 259–284. ISSN: 15427633. DOI: 10.1080/13875868.2018.1431646.

Green, Georgia M (1996). *Pragmatics and natural language understanding.* London: Psychology Press.

Guldi, Jo (2019). "Parliament's debates about infrastructure: An exercise in using dynamic topic models to synthesize historical change". In: *Technology and Culture* 60.1, pp. 1–33. DOI: 10.1353/tech.2019.0000.

Hill, Linda L (2009). *Georeferencing: The geographic associations of information.* Cambridge: MIT Press.

Jockers, Matthew L (2013). *Macroanalysis: Digital methods and literary history.* Champaign: University of Illinois Press. DOI: 10 . 5406 / illinois / 9780252037528.001.0001.

Kilgarriff, Adam (2001). "Comparing corpora". In: *International Journal of Corpus Linguistics* 6.1, pp. 97–133. ISSN: 1384-6655. DOI: 10.1075/ijcl.6.1.05kil.

Klie, Jan-Christoph, Michael Bugert, Beto Boullosa, Richard Eckart de Castilho, and Iryna Gurevych (2018). "The INCEpTION platform: Machine-Assisted and knowledge-oriented interactive annotation". en. In: *Proceedings of the 27th International Conference on Computational Linguistics: System Demonstrations.* Event Title: The 27th International Conference on Computational Linguistics (COLING 2018). Santa Fe, USA: Association for Computational Linguistics, pp. 5–9. URL: http://tubiblio.ulb.tu-darmstadt.de/106270/.

Lakoff, George and Mark Johnson (2008). *Metaphors we live by.* Chicago: University of Chicago press.

Landis, J Richard and Gary G Koch (1977). "The measurement of observer agreement for categorical data". In: *Biometrics* 33.1, pp. 159–174. ISSN: 0006341X. DOI: 10.2307/2529310.

Lansdall-Welfare, Thomas, Saatviga Sudhahar, James Thompson, Justin Lewis, Nello Cristianini, Amy Gregor, Boon Low, Toby Atkin-Wright, Malcolm Dobson, and Richard Callison (2017). "Content analysis of 150 years of British periodicals". In: *Proceedings of the National Academy of Sciences of the United States of America* 114.4, E457–E465. ISSN: 10916490. DOI: 10.1073/ pnas.1606380114.

Lotman, Yuri M (1990). *Universe of the mind: A semiotic theory of culture.* London: IB Taurus.

Manning, Christopher D. and Hinrich Schutze (1999). *Foundations of statistical natural language processing.* Cambridge: MIT Press. ISBN: 0262133601. DOI: 10.1145/601858.601867.

Martin, Wallace (1972). "The hermeneutic circle and the art of interpretation". In: *Comparative Literature* 24.2, p. 97. ISSN: 00104124. DOI: 10.2307/ 1769963.

McEnery, Tony and Andrew Wilson (2003). *Corpus linguistics.* Edinburgh: Edinburgh University Press.

Moretti, Franco (2013). *Distant reading*. London: Verso Books.

Pustejovsky, James and Amber Stubbs (2012). *Natural language annotation for machine learning*. Vol. 1. Sebastopol: O'reilly. ISBN: 978-1-449-30666-3.

— (2013). *Natural language annotation for machine learning: A guide to corpus-building for applications*. Sebastopol: O'Reilly Media, Inc.

Sennrich, Rico, Gerold Schneider, Martin Volk, and Martin Warin (2009). "A new hybrid dependency parser for German". In: *Von der Form zur Bedeutung: Texte automatisch verarbeiten/From form to meaning: Processing texts automatically. Proceedings of the Biennial GSCL Conference 2009*, pp. 115–124.

Simon, Rainer, Elton Barker, Leif Isaksen, and Pau de Soto Cañamares (2017). "Linked data annotation without the pointy brackets: Introducing Recogito 2". In: *Journal of Map & Geography Libraries* 13.1, pp. 111–132. DOI: 10.1080/15420353.2017.1307303.

Stock, Kristin, Robert C. Pasley, Zoe Gardner, Paul Brindley, Jeremy Morley, and Claudia Cialone (2013). "Creating a corpus of geospatial natural language". In: *Lecture notes in computer science (including subseries Lecture Notes in Artificial Intelligence and Lecture Notes in Bioinformatics)* 8116 LNCS, pp. 279–298. ISSN: 03029743. DOI: 10.1007/978-3-319-01790-7-16.

Suissa, Omri, Avshalom Elmalech, and Maayan Zhitomirsky-Geffet (2022). "Text analysis using deep neural networks in digital humanities and information science". In: *Journal of the Association for Information Science and Technology* 73.2, pp. 268–287. DOI: 10.1002/asi.24544.

Venturini, Tommaso, Nicolas Baya Laffite, Jean-Philippe Cointet, Ian Gray, Vinciane Zabban, and Kari De Pryck (2014). "Three maps and three misunderstandings: A digital mapping of climate diplomacy". In: *Big Data & Society* 1.2, p. 2053951714543804. ISSN: 2053-9517. DOI: 10 . 1177 / 2053951714543804.

Zimmer, Michael (2018). "Addressing conceptual gaps in big data research ethics: An application of contextual integrity". In: *Social Media+ Society* 4.2, p. 2056305118768300. DOI: 10.1177/2056305118768300.

PART II

Case Studies

CHAPTER 4

Showing the Potential of Computational Analysis to Support Environmental Narrative Research

Ross S. Purves

Department of Geography; URPP Language and Space, University of Zurich, Switzerland

Olga Koblet

Department of Geography, University of Zurich, Switzerland

Benjamin Adams

Department of Computer Science and Software Engineering, University of Canterbury, Christchurch, New Zealand

In the second half of this book, we present seven case studies where multidisciplinary research groups explore research questions by examining environmental narratives. Six of the case studies were the result of the workshop organised to initiate the process of writing this book. At the workshop, groups of authors with diverse backgrounds came together and identified both research questions and potential methods to address these questions. These case studies exemplify how research questions, such as those introduced in Chapter 2, can be integrated

How to cite this book chapter:
Purves, Ross S., Olga Koblet, and Benjamin Adams (2022). "Showing the Potential of Computational Analysis to Support Environmental Narrative Research." In: *Unlocking Environmental Narratives: Towards Understanding Human Environment Interactions through Computational Text Analysis.* Ed. by Ross S. Purves, Olga Koblet, and Benjamin Adams. London: Ubiquity Press, pp. 87–92. DOI: https://doi.org/10.5334/bcs.d. License: CC-BY 4.0

with computational methods, as introduced in Chapter 3. The final case study was carried out by a single author, a Masters student participating in a course on geographical analysis of text.

All of the studies described in the subsequent chapters combine micro- and macro-readings of text, with a wide range of computational text analysis methods being used in concert with theory and insights from different disciplinary backgrounds. Methodologically, the resulting back and forth between computational methods and more detailed interpretation can be seen as an example of the hermeneutic circle, where the (re)interpretation of material can occur at different stages of the research. It can manifest itself anywhere, for example in identifying collections and queries to create corpora, in the generation of hypotheses, and most obviously in the analysis and interpretation of results to identify supporting evidence in answering the research questions. The goal of these exemplars is thus not to introduce a single recipe for the computational analysis of texts describing the environment. Rather, it is to show, by example, how the questions that we can ask are dependent on the domain of interest, availability and access to appropriate data, the match between data and computational tools, and not least the methodological and thematic expertise of the research team working on a particular problem.

4.1   Case Study Summaries

Glacial narratives: How can they be captured?

Glaciers are prominent geographic features that have infused popular imagination in different ways over time. Where once they were seen as threats to livelihood, in contemporary times they are a symbol of what is being lost due to climate change while also tied to recreation and tourism. In our first contribution, Katrín Anna Lund, Ludovic Moncla and Gabriel Viehhauser look at how narrative writing about glaciers has changed over time through an investigation of texts about glaciers derived from three sources: articles from *Der Spiegel* newspaper, debates from the British parliament and an existing corpus focused on mountain-related texts, Text+Berg. After filtering the documents by matching different keyword forms of the word 'glacier' in English and German, they use a statistical test to examine how discourse has changed over time and find some evidence that supports the hypothesis, particularly with respect to language discussing the disappearance of glaciers. They also highlight some important limitations in the computational approach, for example, through the use of glacier as a metaphor.

Contributions: Lund did the literature review and wrote the first part of the chapter. Viehhauser and Moncla collected the data, analysed and wrote about the experiments and added to the draft. In collaboration, Lund and Viehhauser wrote the conclusions.

Greening a Post-Industrial City: Applying keyword extractor methods to monitor a fast-changing environmental narrative

As we saw already with narratives about glaciers, the ways in which people write about environments reflects not only the physical features of places but also the social context from which the writing stems. In our second contribution, Sarah Luria and Ricardo Campos ask the question of whether an unsupervised keyword extraction tool called YAKE! can give new insight into the complex discourse that surrounds 'revitalization' efforts in post-industrial urban areas. In this study, text documents are collected to give a diversity of perspectives on a specific location, the Canal District of Worcester, Massachusetts, and then keywords are extracted. The automatically generated keywords for the texts are then compared against a close reading interpretation, which shows in what ways the method can help to support environmental history. The authors also highlight some lessons learned from working in an interdisciplinary way.

Contributions: Luria provided the initial question, framing and data set for the chapter. Campos pre-processed all the documents collected by Luria, from text files, images, word documents and pdfs (through Optical Character Recognition [OCR]). Campos processed all of the data through YAKE and produced the results as tables and word clouds. Campos elaborated a Python notebook to allow for the reproducibility of the results. Campos gave guidance on how to read YAKE's results, which Luria then analysed and interpreted. Luria led the writing of the essay and its revision with Campos writing the sections pertaining to YAKE and its functions, and offering comments on the essay as a whole.

Best Practice for Forensic Fishing: Combining text processing with an environmental history view of historic travel writing in Loch Lomond, Scotland

In Chapter 5, Karen Jones, Diana Maynard and Flurina Wartmann, explore the use of text processing on a corpus of travel writing that is geographically and temporally focused on Loch Lomond in the 18th and 19th centuries in order to do historical research on the origin of the region as a touristic landscape. In contrast to the previous two chapters which focused on combining a single computational method with close reading, the approach used here is to consider a number of different computational text analysis methods that are built into the General Architecture for Text Engineering (GATE) toolkit. Tools for information extraction from unstructured text, such as GATE, are now ubiquitous in a plethora of disciplines, domains and applications, and there are numerous freely available open-source possibilities, of which GATE[1], Stanford CoreNLP[2] and NLTK[3] are among the most popular due to their ease of use and adaptability. The authors find that the computational methods provide insight into the use of terminology in historical environmental texts, yet close reading is required in order to provide background information that is 'hidden' in the text and inaccessible to methods that only analyse the surface text.

Contributions: The authors collaborated equally on the scoping of ideas for the chapter, the proposal for integration of different methodologies, and the mechanisms for practical implementation of the study. Maynard and Wartmann led the data assembly, analysis and interpretation of the data, comprising the main body of the chapter. Their findings were calibrated by Jones against a traditional historical/archival interpretation, driving the discussion of challenges in the conclusion, which was written collectively.

The wild process: Constructing multi-scalar environmental narratives

An important class of environmental writing centers on the personal, experiential process of the author interacting with, traveling through, or remembering an environment. Because these narratives present the author's individual experience, who the author is provides additional context and meaning for researchers to be able to analyse the texts. In this chapter, Joanna Taylor and Benjamin Adams analyse environmental texts through the lens of male and female genders. The goal is to test the hypothesis, put forward by Kathleen Jamie, that popular notions of 'wildness' in environmental writing represents a particularly male perspective. Two corpora are constructed: a small set of documents by male and female authors about Rannoch Moor, and a larger corpus of articles from the *Guardian* newspaper's Country Diary column. They employ a multiscalar approach that uses collocation and concordance analysis as well as supervised classification to give insight into general trends in the language in both corpora. Close reading of individual texts provides additional understanding to what the computational results mean allowing them to evaluate the results as evidence for the original hypothesis.

Contributions: Adams collected the data for the Guardian Country Diary and led the computational analysis of that collection. Taylor provided the theoretical approaches and close readings of the texts, and led on the chapter's writing.

---

[1] http://gate.ac.uk
[2] https://stanfordnlp.github.io/CoreNLP/
[3] https://www.nltk.org/

Inferring Value: A Multiscalar Analysis of Landscape Character Assessments

Another domain of environmental writing relates to public policy. Though these documents might not be written from a first-person perspective, social and political contexts still play a role in how they are written. Meladel Mistica, Joanna E. Taylor, Graham Fairclough and Tim Baldwin create a corpus of landscape character assessments – policy documents that are designed to capture the value of landscapes – and explore how topic modelling in combination with close reading can help to untangle the inherent assumptions about 'value' adopted in different character assessments. Because the documents that comprise the corpus come from a variety of sources and use different typesetting and design, the first part of the chapter explores a set of automated tools for extracting the texts into a form suitable for computational analysis.

Contributions: Taylor led the writing of this chapter, and also the analysis in the close reading and discussion sections. Mistica was responsible for implementing the computational elements of the chapter, including identifying and testing the PDF parsers, accessing the data and running the topic models. Baldwin led the design and evaluation of the computational analysis. Fairclough's knowledge about landscape value and Landscape Character Assessments (LCAs) are evident throughout the chapter, with especial contributions to the introduction.

Interpreting natural spatial language in a fictional text: Analysing natural and urban landscapes in Mary Shelley's Frankenstein

Computational methods for distant reading are often touted for their ability to uncover patterns in large corpora, but they can also be useful for understanding the statistics of language use within single texts. In Chapter 5, Tobias Zuerrer uses concordance analysis on the text of Mary Shelley's *Frankenstein* to test if there is a dichotomy between how seed terms denoting natural and urban geographic features are described in this Romantic era novel. In a fictional text, the perspective of a character or narrator might be different from the perspective of the author, which adds additional complexity to the analysis.

Discovering spatial referencing strategies in environmental narratives

In the previous six case studies, existing computational methods are applied along with more traditional qualitative analyses to varying success to answer a number of research questions. The last case study by Simon Scheider, Ludovic Moncla and Gabriel Viehhauser puts the focus on the development of a new computational method for analysing spatial language in environmental narratives that aligns closer with how people visually conceptualise what they read. In particular, they propose a method to explicitly capture spatial frames of reference that are described in natural language. In this case, the two mountaineering texts that are selected are annotated by human readers and then are used to evaluate the efficacy of the new method. The challenges posed in this chapter are indicative of some of the hurdles that researchers face when developing automated natural language processing tools that can approximate some of the more nuanced, close reading tasks performed by domain experts.

Contributions: Scheider contributed the introduction and the parts about spatial reference frames. Viehhauser designed the method for measuring inter-annotator agreement, analysed the results and contributed linguistic background theory. Moncla iteratively designed the parsing rules and ran parsers over all text sources. All authors annotated some texts (in addition to the non-author annotators) and wrote the results discussion together.

CHAPTER 5

Glacial Narratives: How Can They Be Captured?

Katrín Anna Lund
Institute of Life and Environmental Sciences, University of Iceland, Iceland

Ludovic Moncla
Univ Lyon, INSA Lyon, CNRS, UCBL, LIRIS, UMR5205, F-69622, France

Gabriel Viehhauser
Department of Digital Humanities, University of Stuttgart, Germany

On 18th of August 2019, a funeral ceremony was held in the Highlands of Iceland at which the glacier Ok was commemorated, as the first in Iceland to disappear. The ceremony was initiated by two anthropologists from Rice University, Texas, Cymene Howe and Dominic Boyer, both of whom have researched climate change in the Anthropocene. Joining them were about 100 people who followed in a procession to the glacier's former location, including the prime minister of Iceland, Katrín Jakobsdóttir, the former UN human rights commissioner and the president of Ireland, Mary Robinson, glaciologist, Oddur Sigurðsson and writer and environmental activist, Andri Snær Magnason, who authored the text 'A letter to the future' written on the memorial plaque:

How to cite this book chapter:
Lund, Katrín A., Ludovic Moncla, and Gabriel Viehhauser (2022). "Glacial Narratives: How Can They Be Captured?" In: *Unlocking Environmental Narratives: Towards Understanding Human Environment Interactions through Computational Text Analysis*. Ed. by Ross S. Purves, Olga Koblet, and Benjamin Adams. London: Ubiquity Press, pp. 93–108. DOI: https://doi.org/10.5334/bcs.e. License: CC-BY 4.0

Ok is the first Icelandic glacier to lose its status as a glacier. In the next 200 years all our glaciers are expected to follow the same path. This moment is to acknowledge that we know what is happening and needs to be done. Only you know we did.

Ok has officially been declared extinct but the narratives continue so that we will remember and learn from it and be reminded about the consequences of global warming.

Going back to my time in geography classes at primary school, learning about Icelandic nature, I remember Ok being mentioned as the smallest glacier in Iceland that would probably disappear soon; no emotions attached. This was a fact of life, Ok was a small glacier that would not survive warmer conditions; no risk attached. The funeral reflects how the discourse regarding global warming and glacier melting has changed, with a variety of emotions attached to a future at risk, confirmed by melting glaciers.

At school in Iceland, 40 years ago we learned about the main features of nature: glaciers, mountains, rivers, waterfalls and fjords. We learned their names, how to locate them and their utilitarian and aesthetic values. But we did not learn to listen to the stories that nature tells. Rather, we learned to perceive landscape from a visual perspective, locating features as points on a two-dimensional map. Today, people want to engage more closely with nature, it still has its forms, features, names and locations, but, in addition, it has been given a voice. Still, it is a rather passive voice, because it is humans who select the narratives they want to hear by reading into signs stemming from nature. In those narratives glaciers are interesting, they have become, as pointed out by environmental historian, Mark Carey (2007), an 'endangered species'. They are a symbol for climate change. Glaciers are retreating, a warning sign hinting at what may happen if we, the humans, at least in the Western world, continue to treat nature as we have done until now. As storytellers for scientific knowledge, glaciers accumulate a history of changing climate and natural conditions, hence, their important role as natural laboratories. Simultaneously though, they contain narratives of encounters with their human neighbors and how their movements and narratives have been interpreted in different contexts, hence they too have a social and cultural history.

The aim of this chapter is to examine the variety of narratives glaciers have told at different times and are still telling. It can be argued that simultaneously they contain narratives that can be a subject of scientific investigations, and for others who want to listen, and interpret their sayings, they also create their own narratives, not the least in terms of how they can be unpredictable as mobile beings. As such they are creatures of their own nature. So how are their narratives to be captured?

5.1   Living Creatures

In the Western world we often tend to think about our surroundings in terms of dichotomies, cultural on the one hand and natural on the other (Descola and Pálsson, 1996). Nature has become an object out there, to be conquered, researched and adored, but also controlled and activated for human needs. As Pálsson (1996) has argued, Westerners channel their relations with nature predominantly through two pathways of thinking, on the one hand, as exotic and distant, and on the other, through dominating it and situating human beings above. These relational patterns continue in today's environmental discourse. Nature is seen as passive but needing protection and, in that context, it has been provided with life and given a voice, nevertheless as a sensitive being. Thus, today's environmentalists' narratives have provided nature with a voice that calls for help. Nature has reached a point of being at risk. This is why news about the funeral of an extinct glacier in Iceland, that 40 years ago was only regarded as rather insignificant in comparison to other bigger glaciers, is international news. In short, glaciers have become an important symbol of climate change and as a result they attract attention.

Tourists flock to get in touch with glaciers before it is too late. In contemporary society, glaciers have become a valuable product for consumption (Lemelin et al., 2010; Furunes and Mykletun, 2012). As Furunes and Mykletun pointed out, glacial landscapes are viewed as distant and dramatic. They carry the aura of the sublime, that has been

> used to denote vast wilderness with a paradoxical capacity to frighten, but also attract human interest, with thrills and excitement (2013: 327–8).

It appears 19th-century Romanticism is repeating itself but in a different social and political context; the context of a consumer society. Glacier science took off in the 19th century, turning glaciers into natural laboratories. Simultaneously the Romantic era labelled glaciers as 'sublime landscapes and symbols of wilderness' (Brugger et al., 2013, p. 5) and created landscapes of attraction, distant and dramatic enticing 'mountaineers, tourists, and artists seeking awe-inspiring or physically challenging experience' (Ibid.). Science and romanticism created glacier narratives combining feelings for emotions and risk. However, as Douglas and Wildavsky (1983) point out, risk is always selectively defined, depending on geopolitical and social contexts. In the 19th century, scientists dealt with different types of risks from those foreseen by contemporary science. They were working at the end of the era called the Little Ice Age that had been ongoing since the 14th century and glaciers had been advancing. Glaciers were threatening, causing floods, destroying livelihoods and the wilderness they represented was 'remote, desolate, and scary' (Carey, 2007, p. 502).

This was not in the least the situation in the European Alps where, despite their sense of remoteness, people lived in the vicinity of glaciers, although the 'glaciers were mostly in areas were few people ventured' (Wiegandt and Lugon, 2008, p. 34). Glaciers provided water for fertilising land in otherwise harsh mountain surroundings and those who inhabited these surroundings accepted the risk. However, the image provided of glacier environments as remote and perilous also added to their aura as sublime beings, unpredictable and not to be disciplined, thus, the combination of immense beauty evoked through their capacity to frighten.

Whilst people living in European Alpine areas had to deal with increased, unexpected and sometimes disastrous flooding during the era of Little Ice Age co-habitation with glaciers, in Iceland it also became more difficult as they expanded towards the inhabited coastal lowlands. During the first ages of settlement, from around 900, glaciers did not influence the daily life of the inhabitants as they were located in the uninhabited highlands and the glacial rivers streaming over to the coastal areas did not influence movements of people mostly travelling by sea (Björnsson, 2016). However, although at a distance, they evoked mysterious stories about hidden and fertile valleys inhabited by outlaws (Ibid.). In fact, all kinds of myths were associated with glaciers. The most famous one in Iceland is probably the one about Bárður, who was the first settler in Snæfellsnes peninsula on the west coast. The 13th-century story recites how Bárður, half human and half giant, walked and disappeared into the Snæfells glacier at a time he experienced irredeemable grief, never to return. He was for centuries, and still is by some, regarded to be the protector of life in the area. Thus, before glaciers started advancing in Iceland, during Little Ice Age, the risk associated with glaciers was more connected to the mysteries they entailed providing them with the sense for sublimity. However, as glaciers advanced co-habitation became more problematic, and the fact that glacially covered terrain was also volcanic was an additional threat. The interplay of fire and ice evoked narratives of the proximity of hell to earth (Jóhannesdóttir, 2015) that enticed the 'imagination to imagine the enormous powers of the earth that creates this landscape' (Ibid., 61), not the least in the mind of those who traveled from a distance. Glacial activities in the Alpine regions also formed a 'rich body of legends and stories' (Wiegandt and Lugon, 2008) in which disasters were blamed on collective sins.

From the account above it becomes evident how important the context of the encounters between humans and glaciers is in terms of what narratives are and how we, as inhabitants of earth, locate ourselves in relation to glaciers physically and mentally. The question is how we live with them and what do they feature? Before the Little Ice Age in Iceland, glaciers shaped a body of legends related to other worldly creatures, outlaws, humans and giants. On the other hand, when they started advancing and became a threat, the legends hinted at more evil forces at work, invisible, stemming from below the earth or even a moral revenge from forces above. At this time, modern science had not entered the stage and the cosmos was differently ordered. Living with, what today is regarded to be otherworldly, was a part of the everyday (Lund, 2015; Lund and Jóhannesson, 2016), nature and culture had not been separated into two domains as they are now typically perceived in contemporary Western Society. Glaciers were alive and moving and their activities were felt directly. Risk was constantly near. Science, on the other hand made an attempt to control the movements, by finding ways to predict them. Still, glaciers are not to be disciplined, the risk is still inherent.

It is interesting in this context to consider the materiality of glaciers especially given the conventional Western notion of thinking about culture as something separated from nature. This type of dualistic thinking permeates much of our contemporary world view – the Enlightenment taught us how to look at the world as a mosaic, combined by separate units that add up to a whole and that is how we tend to think about landscape, as a surface on which we live, furnished with features and forms (Ingold, 2011). And as Ingold has argued, that is how we conventionally think about landscape, as a surface, of which other less solid things are not a part, such as the sky, weather and the air we breathe. In this context, it can be argued that glaciers are difficult to single out; they are an in-between phenomenon, not solid earth, but at the same time tangible in their materiality as frozen water, but extremely mobile and thus hard to predict. Hence, despite attempts to gain control by predicting glacier behaviour, the sense of risk persists. However, it manifests in a new contextual framework, now one of science and simultaneously awe.

In his writings, Carey (2007) reflects on how the Little Ice Age created a fear that advancing glaciers, especially polar glaciers, could result in ice covering the majority of the earth. Thus, the short period of glaciers advancing

> ...fueled apocalyptic visions of colossal ice sheets descending from the earth's poles to join with mountain glaciers and erase civilisation (Carey, 2007, p. 502).

While scientists worried about the future of humanity in an icy world there were others that attempted to conquer these awesome creatures through different activities. Artists and poets admired them from afar, affected by their sublime characteristics which were expressed through work of arts. At the same time, there were those who went further as they were urged 'to prove their masculinity or femininity, to explore new heights, or challenge their mettle against capricious glaciers' (Carey, 2007, p. 504). Glaciers became fields for recreation. They continue to enthuse scientific, artistic and recreational activities, and do so in the contemporary Western world more than ever before. However, glaciers have ceased to advance and are instead rapidly declining, which has changed how humans live with them. Today they not only put humans and their livelihood at risk, they are at risk themselves as a result of irresponsible behaviour by their human neighbours. Glaciers still represent wilderness and are experienced as distant but simultaneously advanced technology, transport and information, has moved them closer to the everyday life of humans. Increased awareness of climate change and environments at risk, not least through environmental discourses, has made people sense strongly that they live with nature and have responsibilities, although they may not be personally affected (Isenhour, 2010; Brugger et al., 2013).

As a result, people have become more aware of the possibility of vanishing environments, many of whom want to go and experience these evanescing landscapes in person as wildernesses becomes more accessible. Last chance tourism is a new trend in travel encouraging people to travel long distances to witness landscapes in transformation, amongst them glacial landscapes (Lemelin et al., 2010; Furunes and Mykletun, 2012). To approach endangered glacial landscapes is especially popular in the far north and south whilst there is a slightly different story to be told in the European Alpine region. As in Iceland and other northern regions, glaciers have gained importance as recreational places to which tourists are guided, often by local people. So as a source for livelihood they have been gaining a new role in regions where tourism, as an economic activity, has been expanding over recent decades. However, as glaciers melt in Alpine regions ski areas may lose value due to a lack of snow and ice and vanishing glaciers will leave naked mountain landscapes lacking aesthetic appeal (Brugger et al., 2013).

In closing, glaciers not only store valuable information about changing climate and natural conditions. As living creatures that have co-habitated the environment with people for centuries, they also—as pointed out by Brugger, et al. (2013) – store memories. They are cultural and social, as much as natural beings, telling stories about how nature and culture are a patchwork of human and non-human co-habitation, and about how the dynamic of constantly changing relations between different earthly beings shape our cultural/natural environment.

5.2   Experiments

5.2.1   Corpus

In the following, we want to explore how digital methods can help detect narratives told about glaciers in different times and places. We will apply those methods to a macroanalysis (aka distant reading) (Moretti, 2000), tracing stories related to glaciers in text corpora too large for a single researcher to read. Using this approach we aim to observe patterns, shifts or differences in the discourse about glaciers on a large scale. However, since digital distant reading approaches always focus on the 'big picture', these methods can only trace out a broad-brush picture and have to be supplemented with a qualitative close reading.

One of the biggest problems of distant reading analysis is the availability of corpora. Our exploration of glacier narratives is constrained by the breadth of sources available in different languages and spanning different time periods, and as such provides a much shallower account than that outlined in the previous section. Nonetheless, we apply an exploratory approach based on three different, and contrasting, corpora:

1. a corpus of articles from the influential German news magazine *Der Spiegel*;
2. a corpus of debates from the UK parliament, accessed through the website http://www.theyworkforyou.com; and
3. the 'Text+Berg' corpus that features the yearbooks of the 'Schweizer Alpenclub', in German and French (Volk et al., 2010).

These three corpora, we surmised, could potentially contain differing narratives. Glaciers form some of the core landforms likely to be described in 'Text+Berg', and are central to the narrative of mountaineering irrespective of time. In *Der Spiegel* and the parliamentary corpora, however, we expected to see more changes over time, as modern concern and evidence of climate change increases.

For our experiments, we extracted all articles and contributions to British parliamentary debates containing the term 'glacier' (in English) or 'Gletscher' (in German), including alternate forms, such as plurals. After pre-processing, the Spiegel corpus contained 600 articles with a total of 1,187,502 tokens, ranging from the year 1950 to 2015. The distribution of the articles with respect to the tokens per year is shown in Figure 5.1a. The number of articles per year steadily increases until the year 2000, at which it reaches a peak with 28 articles featuring the word 'Gletscher' or an alternate form.

The corpus based on UK Parliamentary debates features 105 contributions and a total of 122,489 tokens. It offers the widest temporal range of all three corpora, spanning the years 1919 to 2020. However, mentions of the word 'glacier' are less common than in the other corpora. The maximum numbers of contributions per year is nine in 2019 (Figure 5.1b).

Finally, from the Text+Berg-Corpus, we used the German versions of the yearbook "Die Alpen", beginning in 1957. This is the first year, in which the yearbook was published in two parallel versions, one in German, the other French. The last issue in the corpus we analysed dates from 2009. In total, 953 articles with 3,360,540 tokens are taken into account. Figure 5.1c shows the distribution of articles that feature forms of the word 'Gletscher' and the distribution of the tokens of these articles. There is a noticeable peak in the number of articles beginning with the year 1996, which however is not reflected in the

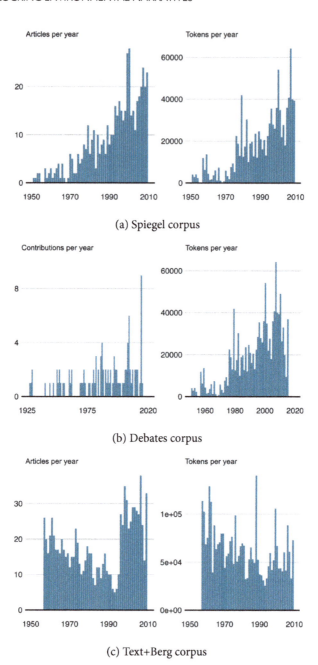

(a) Spiegel corpus

(b) Debates corpus

(c) Text+Berg corpus

Figure 5.1: Articles and tokens per year in the three corpora.

number of tokens. Overall, the articles that deal with glaciers seem to be more equally distributed than in the other two corpora, probably reflecting the specific focus of the journal on mountaineering-related issues.

5.2.2 Word frequencies

For a first look at the content of the corpora, we visualise the most frequent words appearing in the articles using word clouds. Figure 5.2 shows the most common words in each corpora (after removal of stop words). The corpora show quite significant differences, which can be related to the different genres of the three sources. For example, the Text+Berg corpus (Figure 5.2c) focuses on the mountain-related aspects of glaciers (as can be seen in words like 'gipfel'-'peak', 'alpen' - 'alps', and 'schnee' - 'snow' or 'hütte' - 'hut'). It is the only corpus in which the target word ('gletscher' resp. 'glacier') is also the most frequent word in articles containing it. Thus, it seems likely that the articles of the 'Alpen'-yearbook that feature glaciers treat them as a major theme, whereas they are more often only a side aspect in the texts of the other two corpora. Measurements (like 'meter', 'jahr' - 'year', 'uhr' - 'clock', 'zeit' - 'time') also appear in the Text+Berg corpus, but they are less central than in the Spiegel corpus (Figure 5.2a), where 'jahre' ('years') and especially 'jahren' ('years' in oblique case) are – a bit surprisingly – the most frequent words in the whole corpus. This points to a discourse about change, and presumably retreat or disappearance, in the Spiegel corpus, reflecting some of our introductory remarks. Besides measurements, it seems that the Spiegel corpus is also concerned with people and especially Germany in relation to Gletscher ('menschen' – 'human beings', 'deutschen' – 'German'). As can be expected by its genre, the debates corpus (Figure 5.2b) is dominated by government-related words. Remarkably, it is the only corpus that features 'climate' and 'change' as highly frequent words.

5.2.3 Diachronic keyness

A more differentiated picture can be achieved by reflecting on the diachronic change of important words in the corpus. For the experiments in this subsection we first extracted a window of five words before and after the word 'gletscher' resp. 'glacier', to get a better picture of the collocations of our target word in the texts. In a second step we divided the corpora into two parts, one with texts that appear before January 2000 and one with texts appearing in or after 2000. We then calculated the keyness of the words in the younger sub-corpus compared to the older texts using the help of the Chi square measurement (Dunning, 1993) (Figure 5.3).

Interestingly, in both the Spiegel and the debates corpus, the synonymous words 'schmelzend' resp. 'melting' are the most distinctive between the younger

(a) Spiegel corpus

(b) Debates corpus

(c) Text+Berg corpus

Figure 5.2: Word clouds of the most frequent words in the corpora.

and the older texts. Words that describe the vanishing or retreating of glaciers are also amongst the other distinct words in both corpora (in the Spiegel corpus: 'schmelzen' – 'melt', 'verschwunden' – 'vanished', 'schrumpfen' – 'shrink' and also 'kalbenden' – 'calving'; in the debates corpus: 'melt' and 'retreating'). Although not as obvious as in the first two corpora, a similar trend can also be observed in the Text+Berg corpus (besides 'zurückgezogen' – 'retreated' the words 'kalbend' or 'kalbende' – 'calving' are very characteristic for the texts after 2000). Thus, our distant reading approach supports the qualitative assessment that there are changes in how people speak about glaciers in different times. It appears that glaciers are perceived more and more as vanishing, melting or retreating objects in the 21st century.

This gives a glimpse of the possibilities offered by such distant reading methods, which make it possible to carry out analyses, confirm certain hypotheses and identify interesting sub-parts of the corpus that require more attention.

However, there are inevitably cases in these documents where the word glacier is used with a different meaning. This is particularly true for the oldest texts of the Debates corpus as shown in examples (1.) and (2.). Here we note that it is often the case that glaciers are used metaphorically, representing slow, inexorable movement.

1. This is going on like a slowly moving glacier in spite of what is happening abroad and in spite of invitations to Conferences (Debates corpus, 1927).
2. …it has the irresistible movement of a glacier and presents one of our most anxious problems (Debates corpus, 1940).

There are also more unexpected cases, such as the use of glacier as part of a organisational names (3.).

3. …the Glacier Metal Company here in London…(Debates corpus, 1947).

Nonetheless, we also find many mentions of melting glaciers, both in older debates (e.g., 4. from 1929) and many more example from more recent years (see examples 5. to 8.).

4. I should like to take this further opportunity of expressing the deepest sympathy with the victims of the recent floods, which were due to a rain fall 10 times greater in the one month than is normal in the whole year, combined with the bursting of the Shyok Glacier in the Upper Indus Valley (Debates corpus, 1929).
5. Almost 90% of the glaciers have retreated since the 1960s when my father spent two years there with the British Antarctic Survey, but I am

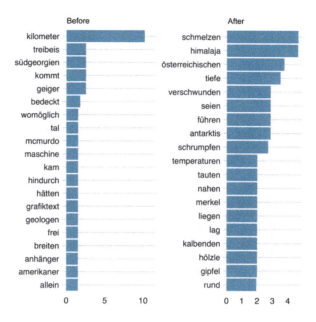

(a) Spiegel corpus

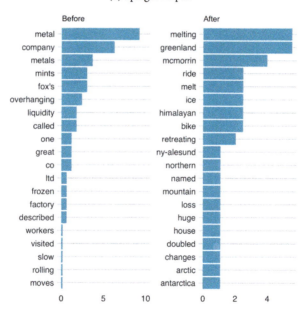

(b) Debates corpus

Figure 5.3: Most distinctive words for texts appearing before and after January 2000.

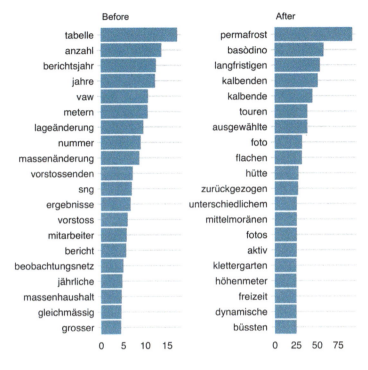

(c) Text+Berg corpus

Figure 5.3: (*continued*).

hopeful that the McMorrin glacier, which was named after him, will still be there when my children are older (Debates corpus, 2019).

6. Glaciers are retreating almost everywhere in the world, from the Alps to the Himalayas (Debates corpus, 2019).

7. Vast cracks have been spotted that could lead to a large part of the glacier breaking away (Debates corpus, 2020).

8. Two thirds of the world's glaciers will have melted, increasing sea levels and drying up rivers across the world (Debates corpus, 2020).

5.3 Discussion

In opening this chapter we set out to explore how glacial narratives can be captured. What became evident was how glaciers' biographical narratives are complex, depending on their changing mobility and how they express their existence regarding their environmental co-habitation. However, their narratives are also selective in terms of what we, as their neighbours, want to hear.

Sometimes we are also forced to listen to what they have to say, creating an aura of risk inherent in their narration, no matter whether they are speaking to us as natural or cultural beings. What is apparent today is that they not only threaten humans, but are also themselves an endangered species. In the era of climate change they have become a symbol for the irresponsible acts towards nature in the Western world. The magnitude of changes to glaciers forces us to listen, and has brought our everyday life into proximity with them. In that context people remind themselves about their destructive behaviour by travelling long distances to get directly in touch with glaciers or even to commemorate those that have already vanished, and by doing that continuing narratives that otherwise might be forever vanish as time passes.

To explore how the narratives that are told about glaciers can be captured, we employed digital methods in an exploratory study described in the second part of our paper. We analysed three different corpora, stemming from different regions and different genres. Our results indicated that a distant reading approach could be helpful as a means to track down traces of the multitude of voices that are hidden in large text corpora and which vary over different times and different places. However, although some of the changes in the discourse related to glaciers can be traced by computational means, we also found limitations of the methods. Firstly, larger digitised and openly accessible corpora are needed to gain a more comprehensive picture and to rule out factors such as genre-specific ways of talking about glaciers. Furthermore, our analyses only uncovered very coarse shifts in discourse, in stark contrast to the subtleties introduced in the first part of the paper. Digging deeper will require more close reading in combination with analysis of richer corpora – for example, recording oral traditions of those living in glaciated parts of the world.

References

Björnsson, Helgi (2016). *The Glaciers of Iceland: A historical, cultural and scientific overview*. Vol. 2. Berlin: Springer. DOI: 10.2991/978-94-6239-207-6.

Brugger, Julie, KW Dunbar, Christine Jurt, and Ben Orlove (2013). "Climates of anxiety: Comparing experience of glacier retreat across three mountain regions". In: *Emotion, Space and Society* 6, pp. 4–13. DOI: 10.1016/j.emospa.2012.05.001.

Carey, Mark (2007). "The history of ice: How glaciers became an endangered species". In: *Environmental History* 12.3, pp. 497–527. DOI: 10.1093/envhis/12.3.497.

Descola, Philippe and Gísli Pálsson (1996). *Nature and society: Anthropological perspectives*. Milton Park: Taylor & Francis.

new wave that's coming in, that's not Worcester', one resident complains. Some see Worcester's development as at a 'tipping point', where Worcester could 'stay Worcester' or be fundamentally redefined, as has happened in places such as Harlem, New York and San Francisco (Schacter, Aaron, 2018b). Such voices can get buried in the attractive 'renaissance' rhetoric of up-scale 'green development'. We hope that by being able to process a range of documents and sources about Worcester, we can produce a more representative picture of public discourse at this critical time.

6.1   Theoretical Approach

Recently economist Robert J. Shiller has argued for the importance of 'economic narratives' and the study of their 'powerful stories' that spread quickly and can influence market behaviours and 'real estate booms' (Shiller, Robert J., 2019b). Worcester offers a powerful example of an appealing economic narrative – something like a phoenix rising from the ashes – of a beleaguered industrial city being 'reclaimed'. Shiller stresses that economic narratives become 'contagious' and are often helped by being promoted by a celebrity (e.g., Ronald Reagan's promotion of Reaganomics) (Shiller, Robert J., 2019a). Such dynamics help demystify why some narratives may get heard more than others in the public debate about revitalisation. Furthermore, economic narratives are marketed by particular buzz words, as Neil Smith showed in his seminal study of the rise of gentrification after the 1960s. Smith tracked how, concerned by the increasingly negative connotations of 'gentrification', savvy real-estate developers appropriated the 'language of revitalization, recycling, upgrading, and renaissance' to build support for upscale development (Smith, 1996). Language, as geographers would put it, 'makes place'. Crucially, Sharon Zukin has shown how the success of an economic narrative can be due not to its broad-based appeal but to the conscious agenda of the local power elite and the media outlets that support them. Zukin studied how New York City's elite achieved the conversion of factories into loft apartments in order to create a more high-end real estate in Manhattan. She points to the strategic role the *New York Times* played through its aggressively positive reporting on that trend (Zukin, 1982). Indeed, the conventional media coverage of Worcester's revitalisation today often reads more like a boosterish advertisement than reporting (*Booming Worcester Real Estate* 2019; *Why Worcester Works* 2019).

   Shiller, Smith and Zukin make it clear that if we are to be critical readers of the de-industrialising of cities towards green urban development, we must tune our ears to the narratives that guide it, the discourse that triggers it and the forces that shape it. Shiller counsels us to be on the lookout for narratives that are becoming 'contagious' in today's Canal District. Smith teaches us to be on the lookout for red-flag keywords such as 'revitalization', and 'renaissance' and to keep asking the question just what 'revitalization' means, and for whom? Zukin warns us to track just which sources and personalities dominate the re-development discussion and suggests we try to level the playing field of public discourse by giving more attention to other less powerful local views. How does Worcester talk about itself? Many investors and future residents are coming from outside of Worcester and may encounter what Boston's newspapers or National Public Radio say about the city, rather than the city's main newspaper *The Worcester Telegram Gazette*, and other important local sources. In hopes of making this complex narrative more accessible, we explore the power of computational analysis to digest the copy generated by such a hot debate.

To accomplish this objective, we aim to apply keyword extractor algorithms. The problem of identifying keywords to track narratives within texts is long-standing (Meehan, 1976), but only recently has attracted more attention from computational linguistics (Vossen, Caselli, and Kontzopoulou, 2015; Campos et al., 2018b). With so much information made available online, getting insightful knowledge from unstructured clinical documents (Conway et al., 2019) or news articles (Martinez-Alvarez et al., 2016), to name but a few is now strictly dependent on algorithms to automate this process and reduce the effort of doing this manually. Recent advances in natural language processing (NLP) have made it possible to extract, summarise and create narratives from texts more easily than ever before (Jorge et al., 2019a; Jorge et al., 2019b). Several diverse ways of representing the overall idea of a text or group of texts exist, ranging from TF-IDF and topic modelling as introduced in Chapter 3 through to visualisation approaches, including keyword clouds (Martinez-Alvarez et al., 2016), visual storytelling (Jorge et al., 2019a), and timeline summarisation (McCreadie et al., 2018; Pasquali et al., 2019). Extracting relevant keywords from texts may be one such potential solution. In this work, we aim to apply YAKE! keyword extractor[1] to a set of texts about Worcester's Canal District to see if this tool can help identify the most important topics and keywords of the input text, without actually having to read the whole document, which, even in the case of a small story like the development of one neighborhood, becomes less and less feasible, due to the vast amount of information that is available today (both digitised historic material and current digital media).

6.2   Sources

A wide survey of documents about the Canal District was made using online search engines, including archive.org (digitised historical documents), Academic Search Premier, Nexis Uni, Proquest, and the digital archives of the *New York Times, Boston Globe, Worcester Telegram Gazette (WTG)* and *MassLive.com*. Google searches were useful for identifying other Worcester media sources such as the Worcester's alternative digital newspaper *InCity Times*; and its local magazine's *Vitality, Worcester Magazine,* and *Worcester Business Journal.* Our search was helped by Sarah Luria having already been

---

[1]   http://yake.inesctec.pt

engaged in researching the Canal District's development history and uncovered sources.

For the purpose of this experiment, we limited our corpus to 26 English-language texts that describe the Canal District over time. We felt 26 texts was a sufficient sampling to analyse and still be able, with some confidence, to make conclusions about YAKE!'s efficacy from one humanist user's point of view and yield some interesting results. If we were to continue the experiment, the corpus could be much larger. The shortest text of our curated corpus contained 72 tokens and the longest one, 4165 tokens. Titles of texts, which often summarise its argument, were included in our extracts. Texts were selected to represent some important features of the history of the District, including the creation of the Blackstone Canal, descriptions of its Irish working-class neighborhood, the city's postindustrial decline, and some of the stages in its efforts toward revitalisation. The majority of the texts are from 2018–2019, an intense period of revitalisation efforts, but examples from 1862, 1917, the 1980s and 1990s provide some historic range. A range of voices was sought and include past and current residents, renters and property owners, city leaders (mayor, city manager, city councilors, community activists), and local and outside developers. While most come from the major local newspaper *The Worcester Telegram Gazette*, the corpus includes articles from *InCity Times*, *Vitality*, *Worcester Magazine*, *Worcester Business Journal*, the *New York Times*, *Boston Globe* and *National Public Radio*. Included in this corpus is an excerpt from an acclaimed historical study, which describes the Irish kitchen barrooms of the Canal District, and a 1984 poem by Worcester-born Mary Fell[2], which describes the neighborhood during its decline.

6.3   Method

Responding to the need to deal with today's abundance of information, researchers have increasingly resorted to computer science as a means to extract, understand and create meaningful stories from large samplings of texts. This is easiest in digitally born documents, where the data is directly available ready to be processed, but offers some additional challenges in analog texts, as is the case of this project, which includes scanned historical documents not captured as plain text. Figure 6.1 shows a sample of one of our texts.

In our corpus, texts were in five different formats, including images, PDFs with images, PDFs with plain text, MS Word documents and text documents. Extracting information from the first two types implies a pre-processing stage that involves the use of Optical Character Recognition (OCR), a machine learning technique used to transform images that contain text (e.g., old text that has been scanned, handwritten, typed, etc.) into text itself. In these cases, we resorted to *tesseract*, an open-source OCR package developed by Google, and to

---

[2] Stanzas from the poem "The Prophecy" by Mary Fell have been used with permission of the author. All rights reserved for all elements of the poem.

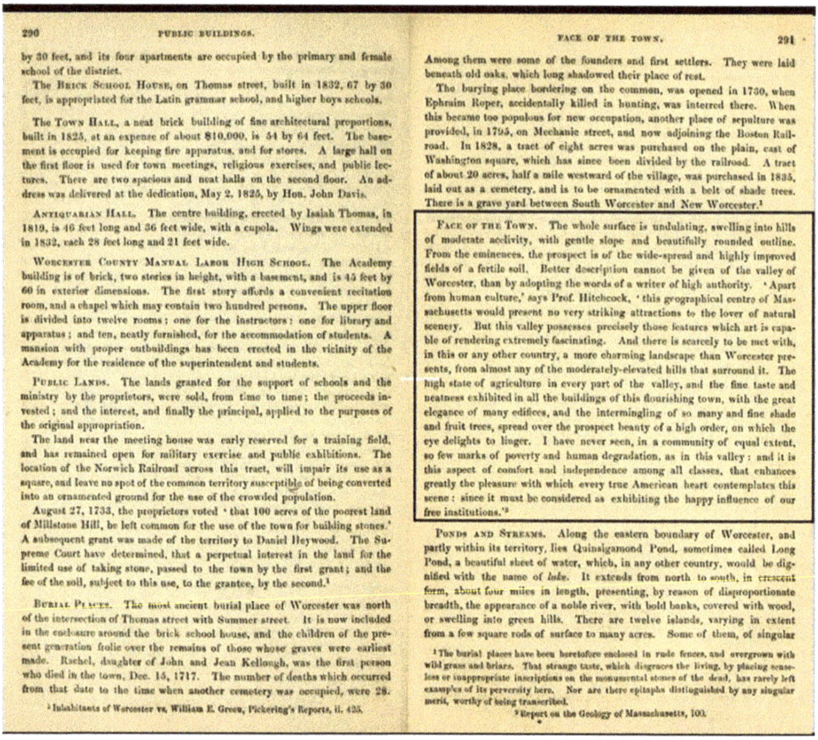

290   PUBLIC BUILDINGS.

by 30 feet, and its four apartments are occupied by the primary and female school of the district.

The BRICK SCHOOL HOUSE, on Thomas street, built in 1832, 67 by 30 feet, is appropriated for the Latin grammar school, and higher boys schools.

The TOWN HALL, a neat brick building of fine architectural proportions, built in 1825, at an expense of about $10,000, is 54 by 64 feet. The basement is occupied for keeping fire apparatus, and for stores. A large hall on the first floor is used for town meetings, religious exercises, and public lectures. There are two spacious and neat halls on the second floor. An address was delivered at the dedication, May 2, 1825, by Hon. John Davis.

ANTIQUARIAN HALL. The centre building, erected by Isaiah Thomas, in 1819, is 46 feet long and 36 feet wide, with a cupola. Wings were extended in 1832, each 28 feet long and 21 feet wide.

WORCESTER COUNTY MANUAL LABOR HIGH SCHOOL. The Academy building is of brick, two stories in height, with a basement, and is 45 feet by 60 in exterior dimensions. The first story affords a convenient recitation room, and a chapel which may contain two hundred persons. The upper floor is divided into twelve rooms; one for the instructors; one for library and apparatus; and ten, neatly furnished, for the accommodation of students. A mansion with proper outbuildings have been erected in the vicinity of the Academy for the residence of the superintendent and students.

PUBLIC LANDS. The lands granted for the support of schools and the ministry by the proprietors, were sold, from time to time; the proceeds invested; and the interest, and finally the principal, applied to the purposes of the original appropriation.

The land near the meeting house was early reserved for a training field, and has remained open for military exercise and public exhibitions. The location of the Norwich Railroad across this tract, will impair its use as a square, and leave no spot of the common territory susceptible of being converted into an ornamented ground for the use of the crowded population.

August 27, 1735, the proprietors voted "that 100 acres of the poorest land" of Millstone Hill, be left common for the use of the town for building stones." A subsequent grant was made of the territory to Daniel Heywood. The Supreme Court have determined, that a perpetual interest in the land for the limited use of taking stone, passed to the first grant; and the fee of the soil, subject to this use, to the grantee, by the second.[1]

BURIAL PLACES. The most ancient burial place of Worcester was north of the intersection of Thomas street with Summer street. It is now included in the enclosure around the brick school house, and the children of the present generation frolic over the remains of those whose graves were earliest made. Rachel, daughter of John and Jean Kellough, was the first person who died in the town, Dec. 15, 1717. The number of deaths which occurred from that date to the time when another cemetery was occupied, were 28.

[1] Inhabitants of Worcester vs. William E. Green, Pickering's Reports, ii. 435.

FACE OF THE TOWN.   291

Among them were some of the founders and first settlers. They were laid beneath old oaks, which long shadowed their place of rest.

The burying place bordering on the common, was opened in 1750, when Ephraim Roper, accidentally killed in hunting, was interred there. When this became too populous for new occupation, another place of sepulture was provided, in 1795, on Mechanic street, and now adjoining the Boston Railroad. In 1828, a tract of eight acres was purchased on the plain, east of Washington square, which has since been divided by the railroad. A tract of about 20 acres, half a mile westward of the village, was purchased in 1835, laid out as a cemetery, and is to be ornamented with a belt of shade trees. There is a grave yard between South Worcester and New Worcester.[1]

FACE OF THE TOWN. The whole surface is undulating, swelling into hills of moderate acclivity, with gentle slope and beautifully rounded outline. From the eminences, the prospect is of the wide-spread and highly improved fields of a fertile soil. Better description cannot be given of the valley of Worcester, than by adopting the words of a writer of high authority. "Apart from human culture," says Prof. Hitchcock, "this geographical centre of Massachusetts would present no very striking attractions to the lover of natural scenery. But this valley possesses precisely those features which art is capable of rendering extremely fascinating. And there is scarcely to be met with, in this or any other country, a more charming landscape than Worcester presents; since almost any of the moderately-elevated hills that surround it. The high state of agriculture in every part of the valley, and the fine taste and neatness exhibited in all the buildings of this flourishing town, with the great elegance of many edifices, and the intermingling of so many and fine shade and fruit trees, spread over the prospect beauty of a high order, on which the eye delights to linger. I have never seen, in a community of equal extent, so few marks of poverty and human degradation, as in this valley; and it is this aspect of comfort and independence among all classes, that enhances greatly the pleasure with which every true American heart contemplates this scene; since it must be considered as exhibiting the happy influence of our free institutions."[2]

PONDS AND STREAMS. Along the eastern boundary of Worcester, and partly within its territory, lies Quinsigamond Pond, sometimes called Long Pond, a beautiful sheet of water, which, in any other country, would be dignified with the name of lake. It extends from north to south, in crescent form, about four miles in length, presenting, by reason of disproportionate breadth, the appearance of a noble river, with bold banks, covered with wood, or swelling into green hills. There are twelve islands, varying in extent from a few square rods of surface to many acres. Some of them, of singular

[1] The burial places have been heretofore enclosed in rude fences, and overgrown with wild grass and briars. That strange taste, which disgraces the living, by placing senseless or inappropriate inscriptions on the monumental stones of the dead, has rarely left examples of its perversity here. Nor are there epitaphs distinguished by any singular merit, worthy of being transcribed.
[2] Report on the Geology of Massachusetts, 100.

Figure 6.1: Lincoln, William. History of Worcester, Massachusetts, from its earliest settlement to September, 1836 : with various notices relating to the history of Worcester County. Worcester, 1862, p. 291. Accessed through archive.org 14 Oct. 2019. Framed text shows material excerpted for corpus selection.

some Python libraries that ease the extraction process. Afterward, we conducted a curated process to remove noisy information and clean-up the text, thus guaranteeing the quality of the data extracted. Then, we applied a keyword extractor system to capture the fundamental idea of the documents. Typically, keyword extractors make available a shortlist of relevant keywords (with one or more terms, and not necessarily the most frequent ones), thus instantaneously providing users and machines with a summary of the document. In an era where most of the information available is unstructured, having one such tool may be very appealing for those interested in quickly getting a sense of a document and in extracting insightful knowledge. Obviously, defining whether a term is a relevant keyword is itself a complex problem that may depend on the use, case, user background or application, such that, as referred by Sterckx et al. (2016), reaching a consensual list of keywords by two different persons for the same document turns out to be a very difficult task. For this purpose, several different techniques have been proposed over the years, from statistical methods that detect keywords based on statistical features (El-Beltagy and Rafea, 2009), to graph-based models (Mihalcea and Tarau, 2004) or machine-learning approaches (McCreadie et al., 2018), a supervised solution that learns from previous examples. A complete survey on keyword extraction algorithms may be found in a recent work of Papagiannopoulou and Tsoumakas (2020).

We grew to realise that for this experiment, Ricardo, a computer scientist, and Sarah, a literature professor, first needed to clarify what each meant by 'keyword'. Indeed, our research went in rather humorously opposite directions until we finally realised the need to back-up and start again by agreeing on definitions of the terms that would guide our research. Note to future interdisciplinary collaborators – define your terms at the outset! Do not assume each of you shares the same ideas about what words mean. To arrive at a shared meaning of 'keyword', Ricardo offered the helpful example of keywords that publishers ask us to assign to our published articles. These words are usually "subject" words that convey the topics of a text. In our corpus, these include names of important places and people in the article, such as 'Canal District', and 'Edward Augustus' (Worcester's City Manager), and important topics, such as 'revitalization', 'gentrification', and 'affordable housing'. Sarah pointed out that since we sought through a keyword extraction algorithm to create a 'summary' of a text, keywords needed to include not just the subject of the article but what was being said about that subject, which she termed the argument, or main point, of the text. Thus, if the article was about the fast rising cost of real estate in the Canal District, keywords would be 'subject' words like 'real estate' and 'Canal District' but also 'argument' words like 'booming' and 'properties are hot' and 'only just beginning'. We thus concluded that for the purpose of this experiment, keywords = subject + argument words.

In this analysis, our chosen tool was YAKE! (Campos et al., 2018a; Campos et al., 2020), an unsupervised statistical keyword extractor method that has demonstrated success in tackling documents from different languages, domains and length, without the need for prior knowledge. Our purpose is to understand whether this kind of algorithm may be used in the context of geography, which focuses on the study of place, to quickly create a flow of stories from a set of documents collected over time and if they help the reader survey the topic being discussed, without the need to acquire further knowledge. With this in mind, we resort to YAKE! Python package[3] to automatically extract the relevant keywords from the set of texts, where a keyword may be a single word or a group of n terms (known as keyphrases).

In this experiment, Sarah was offered five different lists of top-40 relevant YAKE! keywords of our corpus texts with different n settings, namely $n = \{1, 2, 3, 5, 10\}$. Ricardo aimed to offer Sarah the chance to compare the results for

---

[3] https://github.com/LIAAD/yake

the different n's and to see if the effectiveness of their summaries varied to a significant extent in this varied array of texts. She concluded that by and large $n = 5$ and $n = 10$ did not significantly increase the information that $n = 3$ was able to convey about the text. This was a surprise to Sarah, since she had assumed that the more words one extracted, the better the chances that a good summary would be produced. Instead, she discovered that $n = 5$ and $n = 10$ increased the chance for more 'noise' – more little words – that made the summary less clear, and sometimes even inaccurate.

One example illustrates this well. The text is an excerpt from historian Roy Rosenzweig's 1985 study of industrial working-class life in Worcester during its industrial heyday. It describes the kitchen breweries run by Irish women in the Canal District. The excerpt concludes 'It is **unlikely** that these kitchen barrooms were especially lavish or spacious since they shared the physical limitations of most working-class dwellings of this period' (Rosenzweig, 1983). At $n = 5$ YAKE! extracts 'kitchen barrooms were especially lavish', which is the opposite point that the original sentence made. At $n = 10$ YAKE! extracts 'kitchen barrooms were especially lavish or spacious'. At $n = 3$, however, YAKE! extracts the important subject of the 'kitchen grog shops' but does not falsely couple them with the modifiers 'especially lavish [or spacious]'. That said, $n = 3$ does extract (from elsewhere in the passage) the words 'formal and elegant'. The unwary reader could make a false connection in her mind that the 'kitchen grog shops' were 'formal and elegant', which would misrepresent the text"s main point. But it is important that at $n = 3$ YAKE! does not itself falsely connect the two phrases into one keyword phrase as it does at $n = 5$ and $n = 10$.

Based on such results, we opt to define $n = 3$ as our safer and preferred setting, which is in line with the work of Campos et al. (2018, 2020), who pointed out that the most effective results are obtained when the number of grams, that is, the number of terms of a keyword, is set to a maximum of three terms (e.g., 'roads', 'Worcester railroad', 'large manufacturing city').

6.4   Interpretations and Results

After Ricardo processed the corpus texts using YAKE!, the results were then interpreted by Sarah, a literary geographer, who created a gold standard dataset for comparison. Sarah read the original text samples, identified what she considered to be keywords and phrases, and compared them to what YAKE! extracted as an automatic result. Each set of YAKE! extracted keywords was then classified by Sarah as having good, sufficient or insufficient results. In 'good' results, YAKE! extracted enough of Sarah's subject and argument keywords so that the main point of the article is conveyed. In 'sufficient' results, YAKE! extracts some subject keywords and one, or a few, argument words so that the main point of the argument is fairly clear. In 'insufficient' results, YAKE! extracts some subject keywords but no, or not enough, argument key words, so that the argument is not clear.

Good* = 15	Sufficient** = 8	Insufficient*** = 3
1837 Lincoln, History of Worcester[4]	1879 Marvin, History of Worcester County [re Canal and Railroad][5]	1998 Green Island Businesses say city help is killing them
1917 Washburn, Blackstone Canal formed[6]	1985 Rosenzweig, from Eight Hours for What We Will	2000 Green Island revitalization dropped
1917 Washburn, Worcester's entrepreneurial spirit[7]	1999 Vacant Industrial Sites of no use to neighborhood	2019 An Away Game for Businesses[8]
1917 Washburn, importance of steam power[9]	2011 Life in Green Island: We have hope[10]	
1983 Worcester Shedding Smokestack Image[11]	2018 Time to talk gentrification in Worcester[12]	
1984 Fell, The Prophecy[13]	2018 NPR Story Worcester the new It Town[14]	

Table 6.1: Sarah's Evaluation of YAKE! Results – sources linked, where available, online in footnotes.

How many of Sarah's key terms (or close variations of them) did YAKE! catch? Using the above-referred grading scale, Sarah concluded that YAKE!'s summaries of the corpus texts were good in fifteen cases, sufficient in eight and insufficient in three (Table 6.1).

---

4 https://archive.org/details/historyofworcest00inlinc/page/290/mode/2up
5 https://archive.org/details/historyofworcest03marv/page/83/mode/2up
6 https://archive.org/details/historyofworcest03marv/page/83/mode/2up
7 https://archive.org/details/industrialworces00wash/page/300/mode/2up
8 https://www.telegram.com/story/news/local/worcester/2019/07/06/away-game-for-worcester-property-owners-facing-ballpark-redevelopment/4743703007/
9 https://archive.org/details/industrialworces00wash/page/30/mode/2up
10 https://incitytimesworcester.org/tag/millbury-stree/
11 https://www.nytimes.com/1983/09/25/us/worcester-shedding-smokestack-image.html
12 https://www.worcestermag.com/news/20181011/feature-time-to-talk-about-gentrification-in-worcester
13 https://capa.conncoll.edu/fell.persistence.html#38
14 https://www.npr.org/2018/10/23/658263218/forget-oakland-or-hoboken-worcester-mass-is-the-new-it-town

Good* = 15	Sufficient** = 8	Insufficient*** = 3
1989 New focus for an old area[15]	2019 Worcester pledges $3M to Green Island[16]	
1997 Bureau urges liability relief for brownfields	2019 Worcester Organizers hear from Nashville…[17]	
2007 Canal District Shapes Up[18]		
2018 WooSox Ball Park has a long history[19]		
2018 A City Reclaimed		
2019 A Totally Cool Place to Live[20]		
2019 New Shine for old building		
2019 Worcester gets brownfield funds[21]		
2019 Renee Diaz, WooSox killing [Canal] district dreams[22]		

Table 6.1: (*continued*).

---

[15] https://www.nytimes.com/1989/12/10/realestate/national-notebook-worcester-mass-a-new-focus-for-an-old-area.html

[16] https://www.telegram.com/story/news/local/worcester/2019/04/30/worcester-pledges-3m-to-green-island-neighborhood-vows-it-wont-be-overshadowed-by-ballpark/5307782007/

[17] https://www.worcestermag.com/story/news/2019/02/28/worcester-organizers-hear-from-nashville-buffalo-for-tips-on-woosox-cba-push/5802104007/

[18] https://www.telegram.com/story/news/local/east-valley/2007/09/25/canal-district-shapes-up/52786534007/

[19] https://www.bostonglobe.com/business/2018/08/17/new-home-for-woosox-has-long-history/oUqIGARpKusD1NrNItDIaI/story.html

[20] https://www.masslive.com/worcester/2019/04/a-totally-cool-place-to-live-allen-fletcher-offers-sneak-peek-inside-new-kelley-square-lofts-bringing-48-high-end-units-to-worcesters-canal-district.html

[21] https://www.telegram.com/story/news/local/worcester/2019/06/23/worcester-gets-federal-money-for-brownfields-cleanup/4846706007/

[22] https://www.wbjournal.com/article/construction-woosox-regulation-are-killing-canal-district-dreams

These positive results encourage us to think that YAKE! has the potential to serve as an effective summarising tool. Of course, Sarah's "algorithm" or grading rubric was not as precisely or clearly formulated as YAKE!'s; nevertheless, Sarah did find that in the majority of cases YAKE! did extract enough of what she deemed keywords to present a good or sufficient summary of a text. These findings pointed to several primary ways we might hope YAKE! results could be made increasingly informative by tweaking YAKE!s algorithm so it could more closely match Sarah's results. We discuss these briefly below.

As stated above, Sarah generated her list of keywords (subject + argument) for each text in the corpus. Table 6.2 shows one example that Sarah ranked 'good,' which YAKE! generated from Mary Fell's poem 'The Prophecy' (1984). The poem is written from the point of view of a long-term resident in the neighbourhood. Despite the dominant Worcester-narrative of the neighbourhood is in decline, the poem's speaker argues nevertheless that the 'neighborhood

Sarah's Keywords Extracted by YAKE!	Sarah's Keywords Missed by YAKE!
Jews settling	same old bars
Green island	big stories
Remembering the canal	immigrants
Learn polish prayers	
Irish laborers	
Patsy spoke	
Aggie	
Catholic school	
Whiskey	
Aggie brew	
The Neighborhood remains	
Kids	
Built by Irish	
Canal that cut	
Persistence of memory	
prophecy	
Millbury and Harding Streets	
made beer	
American born	
Kelly Square	

Table 6.2: 'Good' Rating: Subject + Argument Words Extracted.

remains'. As Table 6.2 shows, YAKE! extracted almost all of the keywords identified by Sarah.

YAKE!'s results sufficiently capture Fell's emphasis on the ethnic diversity of the canal district– 'jews', 'learn polish prayers', 'irish laborers', 'patsy', 'aggie' (both Irish names), 'catholic school' – as well as her focus on the neighborhood's drinking culture and the 'whiskey', which was 'aggie's brew'. The argument of the poem is also conveyed that 'kids' are still a part of the neighborhood, and 'learn polish prayers', which is in the present tense. A sense of the past is also present through the words 'built by irish', 'remembering the canal', the 'canal that cut' [through this neighborhood], and the 'persistence of memory' (the title of the collection in which the poem appeared). While it is unclear what 'prophecy' means in this context (it is the title of the poem), the word helps convey the weighty, assertive tone of the poem and the suggestion of an (unspecified) future for the neighborhood.

In Sarah's 'sufficient' examples, YAKE! extracted some subject and argument words but omitted keywords critical to the piece. Table 6.3 shows the results for the National Public Radio story 'Forget Oakland or Hoboken, Worcester, Mass. is the New "It" Town' (Schacter, Aaron, 2018a). As the title of that story suggests, this is exactly the sort of discourse that could fan powerful economic narratives that could overwhelm local efforts to achieve a green Worcester for everyone. YAKE!'s words suggest that Worcester is booming, but misses the article's argument that the new real estate boom is just beginning and that local residents are concerned about what this might do for the identity of Worcester.

To see if such minimal summarising happens with other keyword extractors, we ran the very same text under IBM Watson, one of the most well-known commercial solutions. Interestingly, we found that the top-10 keywords retrieved by the IBM system[23] also did not extract the keywords Sarah thought were critical to the piece, demonstrating that there is still much work to do text understanding.

Finally, Table 6.4 shows an example Sarah rated 'Insufficient'. The text is a *Worcester Telegram Gazette* article (7-6-2019) 'An Away Game for Businesses: Property Owners Near Ballpark Make Way for Redevelopment' (*An Away Game for Businesses: Property Owners Near Ballpark Make Way for Redevelopment* 2019). Here YAKE! extracts subject words, but no argument words. As a result, the main point of the article does not come through, which is that tenants have mixed feelings about being displaced.

From Sarah's perspective, two main problems emerged with YAKE!'s results. One was that YAKE! repeats keywords in a text, often several times. In the example from National Public Radio above (Table 6.3), city names (Worcester, San Francisco, Oakland) are repeated and dominate the results (similarly, 'city' was

---

[23] 'smaller cities', 'largest city', 'worst thing', 'expensive city prices people', 'small city', 'cities', 'larger city', 'piece of Worcester', 'City officials' and 'lot of times'.

Sarah's Keywords Extracted by YAKE!	Sarah's Keywords Missed by YAKE!
forget Oakland	New "it" town
Worcester booming industrial	properties are hot
Hoboken	incredibly cheap
beautifying Worcester common	destination for foodies
rise	celebrity chefs
second-largest city	master brewers
expensive city prices	former factories
smaller city	mill buildings
Boston	waiting to die
piece of Worcester	stay away
Massachusetts housing alliance	Union Station
fifth-generation Worcesterite	decrepit and roofless
	only just beginning
	still untapped potential
	sky's the limit
	see everyone succeed
	success begets success
	community spirit
	critical tipping point
	most vulnerable residents
	young families out
	elders out
	crisis
	cash investors
	flip it
	genuinely care about
	maintaining its character
	where immigrants come

Table 6.3: 'Sufficient' Rating: Subject + Some Argument Words Extracted.

repeated seven times out of 10 by IBM Watson's processing of this text, cited above). Sarah would have liked to have seen a list with a greater variety of keywords to create a fuller picture of the text's main point. This is acknowledged by Ricardo as a drawback of YAKE! that deserves further attention in the future. One of the possibilities is to apply a more elaborated deduplication algorithm.

place and its particular situation and challenges. We need to counter the reductive booster narratives of 'green revitalization' generated by the market with more representative, fine-grained accounts to get closer to the truth of what is going on in the streets of these cities (Saunders, 2018). Such diverse local accounts might help undercut the reification of gentrification as an unstoppable economic narrative.

City planning best practice now acknowledges that resident input is essential to any development's success (Myerson, Deborah L., 2004). More and more venues both live and digital are opening up for local residents to speak their needs and visions and play a part in their town's redevelopment[24]. Such input can be strengthened by an increased awareness of the history of the conversation surrounding a neighborhood's development and key terms that have been used to shape it. Such input could also importantly be tracked, and amplified, by tools like YAKE!

Do our results show a more representative and complex picture of what Worcester thinks about 'revitalization'? We think so. Even from this small sampling, our YAKE! word clouds show that city leaders and local residents have been trying to revitalise the city for a long time, in many admirable homegrown ways. Most importantly, they show that the new baseball stadium increasingly dominates Worcester's public conversation about its revitalisation (see Figure 6.3. 2018, *BG*; 2018, *WM*; 2019-2-27 *WTG*, 2019 6-1-*WTG*, 2019-6-20, *WBJ*, 2019-24-6 *WTG*, 2019-24-6 *WTG* 2). Since this article was written the stadium has opened, suggesting that it will take up an increasingly large share of the discursive landscape. Our brief survey suggests too that this will prompt more positive and perhaps increasingly critical discourse from Worcester's residents about the ballpark's development (see Figure 6.3. 2018, *WM*; 6-20-2019, *WBJ*).

Given the insights registered by Shiller, Smith and Zukin, cited above, when it comes to public discourse that fuels economic narratives, we could conclude from our study that some regulation is in order to level the playing field of just who gets heard. Right now, Worcester – its Mayor and City Manager, local press and residents – seem to concur in the desire to 'keep Worcester Worcester'. But public discourse is like a busy traffic intersection. Without a smart traffic light in the center, the largest trucks will succeed in barrelling through, and the local pedestrians and bicyclists, whom many want to encourage, never get a chance to be seen and cross. The Ballpark's development makes good media copy, and it may signal a new phase of Worcester-making that comes, even more than in the past, increasingly from outside rather than local energies. In our furthest reaching reflections from this experiment, we wonder if regulating the flow of discursive traffic through an unsupervised keyword extractor could help Worcester better hear itself think and so navigate its growth through this critical time.

---

[24] CoUrbanize. OnLine Community Engagement. https://www.courbanize.com/

Our turn to an unsupervised keyword extractor in an effort to track more voices is reductive, eclectic and selective, certainly, but we believe these aspects are strengths as well as weaknesses. YAKE! word clouds produce a quick picture, but they also invite one to linger, fathom connections, pay attention to language and survey a discussion's key players and keywords. Thus this project's collaboration may suggest some hopeful first results for one possible approach to create a large engaging canvas of a community conversation that includes voices from the past and present and also highlights the role of language in the ongoing remaking of place. Furthermore, we believe that our collaboration stresses the need for computer scientists and humanists to continue working together to refine an unsupervised keyword extractor like YAKE! to consistently identify significant keywords and produce useful summaries from a wide array of sources on an important question. The potential for such a tool could be significant indeed. In the case of Worcester's Canal District, diversifying the narrative of development seems crucial in the public arbitration over the ethical preservation of history and the greening of place.

Lincoln, History of Worcester, 1862

Washburn, History of Worcester, 1917

Washburn, History of Worcester, 1917

New York Times, 1983

Figure 6.3: YAKE! word cloud (*n* = 3) scroll of corpus.

Mary Fell, 1984

Roy Rosenzweig, 1985

New York Times, 1989

WTG, 1997

WTG, 1999

WTG, 2000

Figure 6.3: (*continued*). YAKE! word cloud (*n* = 3) scroll of corpus.

green island neighborhood
green bamboo restaurant
canal district shapes
walking district lorusso green island
bar dino street
envisions funky green
dino lorusso envisions
building green street district
sleep at night green lorusso bought
knowles loomworks building
millbury street
residents can stroll
restaurant canal

WTG, 2007

community development
community
crompton park life in green
island business
neighborhood families
green island neighborhood
lafayette green island lamartine street
people
park island street work
years life green hope
island community center
blackstone canal
millbury street
blackstone river bikeway

InCity Times, 2011

babe neighborhood
sox aaa affiliate
worcester magazine
named worcester worcesters
polar park worcester team
lots canal district woosox
district association canal red sox
red sox aaa so-called canal
long history years
downtown worcester
blackstone canal

Boston Globe, 2018

past worcester history
downtown
part city forward
change city hall
worcester mayor joseph
events years city lot petty
cities worcester work
worcester city manager
york city people mayor
manager college thing
recreation worcester joe
makes worcester feel
community

Vitality Magazine, 2018

gentrification in worcester
polar park development
community development corporation
red sox officials residential green island
local pawtucket red sox
minor league units neighborhood
development housing van holm
park
gentrification city n't ballpark
percent green island area
polar park worcester green street
union hill green people hill
red sox rents island street
worcester community housing
affordable housing city hall
hill community development
community years
downtown housing developments

Worcester Magazine, 2018

city manager edward
england second-largest city
worcester booming industrial
branca harvard pilgrim health
city prices people branded realty group
nsefast boston housing
hoboken city people
canary
worcester officials worcester boom
worcester city manager
forget oakland oakland
smaller city sargent
massachusetts
san francisco
fifth-generation worcesterite

National Public Radio, 2018

Figure 6.3: (*continued*). YAKE! word cloud (*n* = 3) scroll of corpus.

stand up nashville
benefits sox deal
kelly woosox cba push
buffalo sox community
agreement organizers hear local hiring
people club wednesday night
coalition pni club red sox city augustus jr a
community labor coalition
red worcester red sox barnett
pni club wednesday cba
red sox ballpark nashville
community benefits agreements
offes tips on woosox project
sox ballpark project city manager
successful community benefits
invited john washington
buffalo for tips
manager ed augustus

WTG, 2019-2-27

units facing kelley
facing kelley square
place square
kelley fletcher public market
developer allen fletcher
lofts sneak peek inside
allen allen fletcher offers
kelley square lofts
high-end canal district canal
offers sneak peek
fletcher offers sneak
totally cool place worcester
worcester 's canal
crompton place place to live
high-end apartments

MassLive, 2019

worcester zoning board
chelsea-based developer
street apartment building
street factory shoe water factory on water
converted factory rossi plans
water street building studio
walker shoe factory
show anthony rossi
anthony rossi bought board
studio apartments planning
records show anthony
called walker lofts feet

WTG, 2019-6-1

pawtucket red sox
street massachusetts
employees
queen canal green street
time queen 's cups
regulation are killing
killing canal district
canal district dreams
construction water
business pay work owner
district n't
cups years
worcester

WBJ, 2019-6-20

city square massachusetts democrats
loan fund program
housing money revolving loan fund
agency gateway park project
polar park park project received
talent work epa brownfields phoenix
received brownfields sites sen
environmental protection agency
red sox program city epa park support
edward city manager edward funds
environmental worcester projects
brownfields revolving loan
developing brownfield sites
mcgovern green island island augustus

WTG, 2019-24-6

development block grant
island blvd island
city augustus ballpark
augustus announced monday
residents city officials vow
island community center
green island neighborhood
years green island green
city manager edward
worcester pledges
community development block
city councilor sarai
neighborhood
block grant money

WTG, 2019-24-6

Figure 6.3: (*continued*). YAKE! word cloud (*n* = 3) scroll of corpus.

References

An Away Game for Businesses: Property Owners Near Ballpark Make Way for Redevelopment (2019).

Booming Worcester Real Estate (2019). URL: https://www.wcvb.com/article/booming-worcester-real-estate/25741906#.

Campos, Ricardo, Vítor Mangaravite, Arian Pasquali, Alípio Mário Jorge, Célia Nunes, and Adam Jatowt (2018a). "A text feature based automatic keyword extraction method for single documents". In: *European Conference on Information Retrieval*. Berlin: Springer, pp. 684–691. DOI: 10.1007/978-3-319-76941-7_63.

Campos, Ricardo, Vítor Mangaravite, Arian Pasquali, Alípio Mário Jorge, Célia Nunes, and Adam Jatowt (2018b). "YAKE! collection-independent automatic keyword extractor". In: *Advances in information retrieval*. Vol. 10772. Grenoble: Lecture Notes in Computer Science, pp. 806–810. DOI: 10.1007/978-3-319-76941-7_80.

Campos, Ricardo, Vítor Mangaravite, Arian Pasquali, Alípio Mário Jorge, Celia Nunes, and Adam Jatowt (2020). "YAKE! Keyword extraction from single documents using multiple local features". In: *Information Sciences* 509, pp. 257–289. DOI: 10.1016/j.ins.2019.09.013.

Conway, Mike, Salomeh Keyhani, Lee M. Christensen, Brett R. South, Marzieh Vali, Louise C. Walter, Danielle L. Mowery, Samir E. AbdelRahman, and Wendy W. Chapman (2019). "Moonstone: A novel natural language processing system for inferring social risk from clinical narratives". In: *Journal of Biomedical Semantics* 10. DOI: 10.1186/s13326-019-0198-0.

El-Beltagy, Samhaa R and Ahmed Rafea (2009). "KP-Miner: A keyphrase extraction system for English and Arabic documents". In: *Information Systems* 34.1, pp. 132–144. DOI: 10.1016/j.is.2008.05.002.

Fell, Mary (1984). "The prophecy". In: *The persistence of memory*. New York: Random House.

Jorge, Alípio M, Ricardo Campos, Adam Jatowt, and Sérgio Nunes (2019a). *Information Processing & Management Journal Special Issue on Narrative Extraction from Texts (Text2Story): Preface*. DOI: 10.1016/j.ipm.2019.05.004.

Jorge, Alípio Mário, Ricardo Campos, Adam Jatowt, and Sumit Bhatia (2019b). "Second international workshop on narrative extraction from texts (Text2Story'19)". In: *Advances in Information Retrieval*. Vol. 11438. Cologne: Lecture Notes in Computer Science, pp. 389–393. DOI: 10.1007/978-3-030-15719-7_54.

Martinez-Alvarez, Miguel, Udo Kruschwitz, Gabriella Kazai, Frank Hopfgartner, David Corney, Ricardo Campos, and Dyaa Albakour (2016). "First international workshop on recent trends in news information retrieval (NewsIR'16)". In: *European Conference on Information Retrieval*. Berlin: Springer, pp. 878–882. DOI: 10.1007/978-3-319-30671-1_85.

McCreadie, Richard, Rodrygo LT Santos, Craig Macdonald, and Iadh Ounis (2018). "Explicit diversification of event aspects for temporal summarization". In: *ACM Transactions on Information Systems (TOIS)* 36.3, pp. 1–31. DOI: 10.1145/3158671.

Meehan, J. (1976). *The Metanovel: Writing stories by computer.*

Mihalcea, Rada and Paul Tarau (2004). "Textrank: Bringing order into text". In: *Proceedings of the 2004 Conference on Empirical Methods in Natural Language Processing*, pp. 404–411.

Myerson, Deborah L. (2004). *Involving the Community in Neighborhood Planning*. URL: http : / / uli . org / wp - content / uploads / 2012 / 07 / Report - 1 - Involving-the-Community-in-Neighborhood-Planning.ashx_.pdf.

Papagiannopoulou, Eirini and Grigorios Tsoumakas (2020). "A review of keyphrase extraction". In: *Wiley Interdisciplinary Reviews: Data Mining and Knowledge Discovery* 10.2, e1339. DOI: 10.1002/widm.1339.

Pasquali, Arian, Vítor Mangaravite, Ricardo Campos, Alípio Mário Jorge, and Adam Jatowt (2019). "Interactive system for automatically generating temporal narratives". In: *European Conference on Information Retrieval*. Berlin: Springer, pp. 251–255. DOI: 10.1007/978-3-030-15719-7_34.

Rosenzweig, Roy (1983). *Eight hours for what we will: Workers and leisure in an industrial city 1870-1920*. Cambridge: Cambridge University Press.

Saunders, Pete (2018). "The scales of gentrification". In: *Planning* 84 (11), pp. 16–23. DOI: 10.1044/leader.BGJ.23112018.16.

Schacter, Aaron (2018a). *Forget Oakland or Hoboken, Worcester, Mass. is the New It Town*. https://www.npr.org/2018/10/23/658263218/forget-oakland-or - hoboken - worcester - mass - is - the - new - it - town. [Online; accessed 05-May-2020].

— (2018b). *It's Time to Talk About Gentrification in Worcester*. URL: https : / / www . worcestermag . com / news / 20181011 / feature - time - to - talk - about - gentrification-in-worcester.

Shiller, Robert J. (2019a). *Narrative economics*. Princeton, N.J.: Princeton U Press.

— (2019b). *What People Say about an Economy Can Set of a Recession*. [Online; accessed 05-May-2020]. URL: https : / / www . nytimes . com / 2019 / 09 / 12 / business/recession-fear-talk.html.

Smith, Neil (1996). *The New Urban Frontier: Gentrification and the Revanchist City*. New York: Routledge.

Sterckx, Lucas, Thomas Demeester, Chris Develder, and Cornelia Caragea (2016). "Supervised keyphrase extraction as positive unlabeled learning". In: *EMNLP2016, the Conference on Empirical Methods in Natural Language Processing*, pp. 1–6. DOI: 10.18653/v1/D16-1198.

Vossen, Piek, Tommaso Caselli, and Yiota Kontzopoulou (2015). "Storylines for structuring massive streams of news". In: *Proceedings of the first workshop on computing news storylines*, pp. 40–49. DOI: 10.18653/v1/W15-4507.

Whitmore, Bernard (2018). *A Mayor, A Manager, A City Reclaimed.* [Online; accessed 17-October-2019].

Why Worcester Works (2019). URL: https://www.wcvb.com/article/why-worcester-works/25741937.

Zukin, Sharon, ed. (1982). *Loft living.* Baltimore: Johns Hopkins University Press.

CHAPTER 7

Best Practice for Forensic Fishing: Combining Text Processing with an Environmental History View of Historic Travel Writing in Loch Lomond, Scotland

Karen Jones
School of History, University of Kent, UK

Diana Maynard
Department of Computer Science, University of Sheffield, UK

Flurina Wartmann
Department of Geography & Environment, University of Aberdeen, UK

Scenic landscapes are a main attractor for local and international tourism, and in many cases have become designated as protected areas such as national parks or scenic areas that promote their aesthetic qualities to attract visitors. But what directs touristic attention to certain landscapes, and to specific places within such landscapes? We argue that in order to find out how touristic landscapes come into being, we need to turn our focus on how such landscapes become constructed as idealised landscapes.

How to cite this book chapter:
Jones, Karen, Diana Maynard, and Flurina Wartmann (2022). "Best Practice for Forensic Fishing: Combining Text Processing with an Environmental History View of Historic Travel Writing in Loch Lomond, Scotland." In: *Unlocking Environmental Narratives: Towards Understanding Human Environment Interactions through Computational Text Analysis*. Ed. by Ross S. Purves, Olga Koblet, and Benjamin Adams. London: Ubiquity Press, pp. 133–160. DOI: https://doi.org/10.5334/bcs.g. License: CC-BY 4.0

This chapter is based on the idea that we can trace the construction of landscapes as touristic places through historic text sources, and explore how these texts work in creating a scenic identity of landscape. We take as an illustrative example the Loch Lomond area in Scotland, and examine the 'origin stories' of the scenic sites in accounts of guide/travel books from the late 18th and 19th centuries, exploring the way in which their landscape identities were formulated.

We have developed a hybrid approach that combines qualitative text analysis and interpretation with more quantitative analysis using Natural Language Processing tools. By conducting this parallel 'forensic fishing' activity, we are looking to ascertain two things: firstly, what might be 'missing' from a strict linguistic analysis in terms of historical context or 'buried' information hard to decode from the data alone and, secondly, how the data from a quantitative linguistic analysis might embellish, challenge and/or inform a conventional methodological/archival trawl using environmental humanities techniques. We found that to the environmental historian, these guidebooks show a tourist landscape under construction, crafting a consistent narrative of celebrating the wild and the romantic. This is communicated via a recurring phraseology and a clear choreography of the visitor experience.

The analysis of these sources through the use of automated language processing tools offers insights into the recurring terminology used to describe landscape experience, and allows quantitative comparisons to be made between different sources. However, these methods do not address the issue of concealed messages buried deep in the text, or inferred by illustrations, appendices and supplementary material. These glimpses are important since they offer useful hints as to the hidden stories to be told beyond the dominant narrative about a world of elite travel in Loch Lomond.

In the following, we outline the hybrid approach that we used to trace the emergence of Loch Lomond landscapes as iconic touristic places through historical travel writing. We start by describing our procedure for finding relevant source material and processing this digitised text material to compile a corpus. We then describe the processing steps we took to analyse these texts using the text analysis toolkit 'General Architecture for Text Engineering' known as GATE (Cunningham, Maynard, and Bontcheva, 2011). Using examples, we show how text analysis in GATE can assist and complement a close reading of these texts. Finally, we highlight how different methodological and disciplinary lenses can be applied and combined for examining the origin stories and historic landscape experiences of historic travel writers in Loch Lomond.

7.1   Finding Relevant Sources and Compiling a Corpus of Digitised Travel Descriptions

Data collection is a critical consideration, both in the definition of our research goals and in the design of our experimental study. In order to find relevant material enabling our research questions to be answered, we need to consider the following questions: where can we find digitised historic texts? Are the texts we find biased in terms of coverage; do they cover sufficient material; are they focused on the topic we want to investigate? And finally, do the texts contain relevant material that will allow our questions to be answered?

Our task can be considered as a 'fishing experiment', since we do not necessarily know what we might uncover until we start looking for it. We therefore started with a selection of texts, and used linguistic tools to initially sift through them, enabling us to then zoom in on relevant material with respect to our research question. The sample corpus was selected from the online archives[1] and project Gutenberg[2] using a date-limited search (earliest) of 'Loch Lomond' (text sources only), with a range intended to include different authors and as many guidebooks as possible. The following criteria were thus applied: only text sources were selected (as other source types cannot be processed automatically using our methodology); the 'first' publication date versions of each guidebook were selected; and a date cut-off of 1895 was applied to limit sample size. Using these criteria, we compiled a small corpus consisting of eight digitised works:

1. Journey to the Western Isles, Johnson and Boswell (1785)
    https://www.gutenberg.org/files/2064/2064-h/2064-h.htm
2. The picture of Glasgow, and strangers' guide; with a sketch of a tour to Loch-Lomond, R Chapman (1818)
    https://archive.org/details/pictureofglasgow00unse/page/n9
3. The Steam-boat Companion; and Stranger's Guide to the Western islands and Highlands of Scotland, James Lumsden & Son (1820)
    https://archive.org/details/bub_gb_x5sHAAAAQAAJ/page/n4
4. Guide to the romantic scenery of Loch-Lomond, Loch-Ketturin, the Trosachs, James Lumsden, (1831)
    https://archive.org/details/guidetoromantic00songoog/page/n5
5. Black's Guide to the Trossachs (1853)
    https://archive.org/details/blacksshillinggu00ediniala/page/n4
6. Nelson's Tourist Guide (1858)
    https://archive.org/details/nelsonstouristsg00thom/page/n2

---

[1] http://www.archive.org
[2] http://www.gutenberg.org

7. Edinburgh & Glasgow to Stirling: Doune, Callander, Lake of Menteith, Loch Ard, Loch Achray, the Trosachs, Loch Katrine, Loch Lomond, Keddie and Gray (1873)
https://archive.org/details/edinburghglasgowkedd/page/n13
8. Shearer's Guide to Stirling, Dunblane, Callender, the Trossachs and Loch Lomond, Shearer (1895)
https://archive.org/details/shearersguidetos00shea/page/2

We then compared this selection of sources to British Library holdings and Wikipedia in order to check whether we were missing relevant non-digitised sources. In doing so, we identified several non-digitised works at the British Library, including the Guide to Loch Lomond (Richardson, 1798); Baird's Guide (Baird, 1853), Brydone's Guide (Brydone, 1856); and Shaw's Tourist's Picturesque Guide to the Trosachs, Loch Lomond, Central Highlands (Shaw, 1878). It is important to acknowledge that our corpus is thus a sample only of historic texts of Loch Lomond which are digitised. A more detailed analysis of historic travel writing about the Loch Lomond area should include these non-digitised texts, which could be analysed in an analogous manner or digitised using Optical Character Recognition (OCR) methods, and then be included in the sample. However, for our initial exploratory analysis, we limit ourselves to the available digitised text sources.

7.2   Forensic Fishing in Digital Text Sources Using Natural Language Processing

It is generally very time-consuming for researchers to manually read a number of books and find relevant parts that warrant a closer reading, as well as manually counting the words and their relations to each other. But natural language processing (NLP) approaches can, with some guidance, be used to sift through this large amount of data and quantify the occurrence of certain words or parts of words, which is useful even for a small number of sources. In our approach for text analysis, we use automated tools to find the 'interesting' parts of the documents, which can then be passed to domain experts for manual analysis, and to count the occurrences of descriptive terms that we were interested in that relate to scenicness and aesthetic perception of landscape.

7.2.1   Processing digitised historic texts in GATE to identify landscape terms and place names

Because our corpus is historical, and therefore compiled primarily from text that has been digitised using OCR methods, some initial cleaning of the data was necessary. We manually removed some of the extraneous text denoting advertisements, indexes and so on. We also separated the text into smaller parts, since a whole book as a single document is rather large for processing within the GATE toolkit. This separation can be done manually, or in our case, we used a script that executes a program in Python that automatically separates the single text documents of one book into smaller segments that are easier to analyse.

Once the data was cleaned, we then ran some basic linguistic processing tools through the data collection using GATE. This enabled us to capture words, sentences, paragraphs and parts-of-speech (noun, verb, adjective, etc.). Next, we ran Named Entity Recognition (NER) tools to determine to which categories the identified entities belonged (e.g., whether a noun is a reference to a person, a place name or a landscape feature). We used the default ANNIE NER tool implemented in GATE, which we customised slightly to improve recognition of geographical features, which is not implemented in ANNIE as a default. To do so, we used a list of English terms that describe landscape features. This list contained landscape elements (e.g., *forest, bridge, house, village, river*), qualities that modify these elements (e.g., *beautiful, blue*), as well as activities related to landscape experience (e.g., *walk, walking, hike, hiking, view, viewing*). This list of single words had been compiled from a contemporary data source consisting of social media photographs with associated keywords (tags) in previous research (Purves, Edwardes, and Wood, 2011). As this list reflected contemporary terms, we adapted the original list to our historic corpus by removing words such as 'disco' or 'skyscraper', resulting in a new list of 365 words. We used this list as an initial gazetteer to tag mentions of landscape terms in the historic text data (Annex 1).

abandoned	butterfly	ditch	grassy	mills	reservoir	sunny
abbey	byway	dock	grave	mine	restored	sunrise
access	cairn	docks	gravel	mist	ridge	sunset
agricultural	camp	downs	graves	monument	river	sunshine
air	camping	downstream	gravestone	moon	road	swan
allt	canal	drainage	graveyard	moor	roads	swans
ancient	carriageway	duck	grazing	moorland	roadside	thatched
animal	castle	ducks	grove	moors	robin	tide
animals	cathedral	dyke	hamlet	moss	rock	timber
apple	cattle	edge	harbour	mount	rocks	tomb
arable	cemetery	estuary	haven	mountain	rocky	tourist
arch	cemetery	farm	hay	mountains	roman	tower
arches	chalk	farmhouse	headland	mouth	rough	town
arena	channel	farming	heath	mud	route	track
ash	chapel	farmland	heather	muddy	ruin	tracks

Annex 1: List of landscape terms.

attractive	church	farms	hedge	oak	ruins	traditional
autumn	churches	fauna	heritage	outdoor	rural	trail
bank	churchyard	fell	hiking	outdoors	sailing	train
banks	clay	fen	hill	overgrown	sand	trains
bar	cliff	fence	hills	palace	sands	tree
barn	cliffs	ferry	hillside	panorama	sandstone	trees
barns	climbing	field	historic	park	sandy	tributary
barrow	cloud	fields	history	pass	scenery	upstream
bay	clouds	fire	horizon	pasture	sea	vale
beach	clough	fish	horse	path	seafront	valley
beautiful	coal	fishing	horses	paths	seaside	view
beauty	coast	flood	hut	peak	sheep	viewpoint
beck	coastal	flooded	ice	peat	ship	village
bed	coire	flora	inn	pier	shore	visitors
beech	colliery	flower	insect	pine	signpost	walk
beinn	copse	flowers	island	pit	sky	walkers
ben	cottage	flowing	isle	plant	skyline	walking
bench	cottages	flows	isolated	plantation	slope	wall
bend	countryside	fog	junction	plants	slopes	walls
birch	cove	footbridge	lake	ploughed	snow	water
bird	cow	footpath	landscape	pond	spire	waterfall
birds	crag	footpaths	lane	pool	spring	waves
boat	crags	ford	leaf	port	squirrel	weather
boats	croft	forest	leaves	priory	stars	weir
bog	crop	forestry	lighthouse	quarry	station	well
boggy	cross	fort	limestone	quay	steep	wet
botanical	crossing	frost	loch	quiet	stile	wheat
boundary	crossroads	gap	lochan	railroad	stone	wild
branch	cutting	garden	lock	railway	stones	wildlife
bridge	cwm	gardens	locks	railways	storm	wind
bridges	cycle	gate	lodge	rain	stream	windmill
bridleway	cycling	gates	manor	rainbow	street	winter
brook	dale	gateway	marina	range	streets	wood
building	dam	gill	marsh	rebuilt	summer	wooded
buildings	deer	glen	meadow	reflection	summit	wooden
burn	derelict	gorse	meall	reflections	sun	woodland
busy	disused	grass	mill	reserve	sunlight	woods

Annex 1: *(continued).*

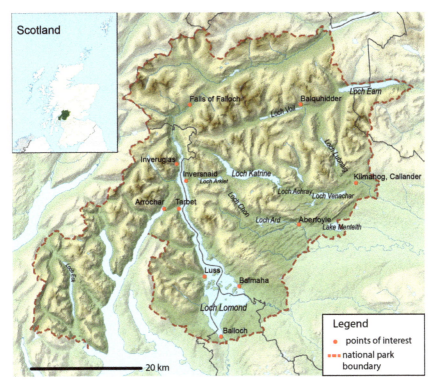

Figure 7.1: Map of Loch Lomond and The Trossachs National Park study area (base map credit: Wikimedia user Nilfanion, map created using Ordnance Survey open data CC-BY-SA-3.0).

Finally, we were interested in identifying sections within these books that specifically talk about landscapes and places in the Loch Lomond area. For this, we used the boundaries of the Loch Lomond and the Trossachs National Park to define our study area, which is rooted in contemporary perception of the National Park as a spatial unit (Figure 7.1).

We recognise that historically, the landscapes that travellers traversed on a journey to the Loch Lomond area also incorporated areas not included in the National Park, such as Stirling, which we excluded from our analysis. We therefore compiled a list of place names for Loch Lomond using the Loch Lomond and The Trossachs National Park homepage to compile a list of 56 place names (settlements and points of interest) as well as a list of 20 names for water bodies (e.g., Loch Lomond, Loch Arklet). We also imported a list of names for Scottish mountains from Wikipedia to recognise references to mountains in Loch Lomond. After we compiled these gazetteers or lists of names (Annex 2), we ran the GATE application over all our texts to identify mentions of these words (Figure 7.2).

Tyndrum	Crianlarich	Stronachlachar	Inversnaid	Ardlui
Inverarnan	Inveruglas	Tarbet	Arrochar	Inverary
Lochgoilhead	Rowardennan	Inverbeg	Garelochhead	Luss
Inchmurrin	Helensburgh	Arden	Balloch	Killin
Balquhidder	Lochearnhead	Strathyre	St. Fillians	Brig o'Turk
Brig O'Turk	Callander	Thornhill	Port of Menteith	Doune
Kinlochard	Aberfoyle	Garmore	Balmaha	Drymen
Gartocharn	Glen Dochart	Duke's Pass	Inchcailloch	Falls of Falloch
Puck's Glen	Falls of Dochart	Bracklinn Falls	Falls of Leny	Falls of Edinample
Edinample Castle	Clarinsh	Inchfad	Inchtavannach	Inchmurrin
Inchlonaig	Inchconnachan	Buchinch	Inchmoan	Inchgalbraith

Annex 2a: List of place names for Loch Lomond and The Trossachs National Park. Settlements and places of interest.

Loch Iubhair	Loch Dochart	Loch Voil	Loch Doine
Loch Venachar	Loch Lubnaig	Loch Lomond	Loch Long
Loch Katrine	Holy Loch	Loch Goil	Loch Earn
Loch Arklet	Lake of Menteith	Loch of Mentieth	Loch Reòidhte
Loch Drunkie	Loch Achray	Loch Ard	Loch Chon

Annex 2b: List of place names for Loch Lomond and The Trossachs National Park. Water bodies.

7.2.2   Identifying relevant texts containing landscape descriptions of the Loch Lomond area

For our analysis, based on the annotations, we first focused on identifying passages containing landscape descriptions of the study area of Loch Lomond by searching for mentions of place names located within the study area that co-occurred with a concentration of landscape terms. Using the number of annotations of landscape terms and Loch Lomond place names in relation to the number of all tokens in a text file allowed us to quickly zoom into text files that were most likely to contain landscape descriptions that were of interest to us. For instance, for the source of Keddy and Gray (1873), this ratio indicates that the text files number 4, 5 and 6 (marked in bold in Table 7.1) contain considerably more relevant terms than others, also when compared to the overall tokens in the text file. These are thus good candidate text files that we want to look at for our more in-depth analysis.

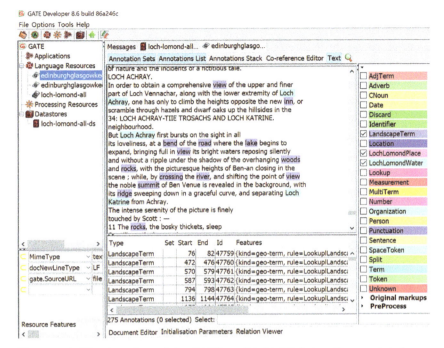

Figure 7.2: User interface of the GATE toolkit highlighting the annotated terms (landscape terms and Loch Lomond place names for settlements and water bodies – see Annex 1 & 2 for full list of search terms used).

However, there may also be passages within those files with lower ratios that might be highly relevant, but which get missed by the first method because those texts do not contain the relevant place names or have only few landscape terms. We therefore perform a second pass manually. Using the GATE toolkit to highlight Loch Lomond place names and landscape terms (Figure 7.2), we visually identified text passages where these co-occur. Taking this approach, we also identified relevant passages, for instance in text file number 3 (Figure 7.2) that contained 21 place names from our study area and 254 landscape terms (Table 7.1).

Using this approach, we compiled a sub-set of documents from each travel guide for more in-depth qualitative analysis ('close reading'). The examples cited below are taken from these text passages highlighted as containing high ratios of relevant place names and landscape terms compared to the overall term count.

We highlight below some examples of approaches that combine a close reading approach of these relevant texts with more quantitative approaches ('distant reading') of the entire corpus. Our examples illustrate how these approaches can be applied and combined to generate initial insights on how the landscapes of Loch Lomond and The Trossachs were experienced and described in historic travel accounts.

Nr.	Number of all place names	Number of Loch Lomond Place Names	Number of landscape terms	Number of tokens in text file	Ratio between landscape terms and Loch Lomond place names to all tokens
1	126	11	100	4277	0.026
2	115	20	234	7047	0.036
3	119	21	254	7402	0.037
4	145	66	402	8220	0.057
5	160	82	244	7272	0.045
6	128	54	269	5668	0.057
7	45	6	98	2581	0.040
8	235	50	100	4957	0.030
9	79	0	12	2602	0.005
10	309	7	16	4040	0.006
11	79	1	13	3822	0.004
12	201	29	39	5511	0.012

Table 7.1: Number of annotated terms identified through gazetteer lookup.

7.2.3 Application example 1: Tracing the emergence of touristic hotspots in Loch Lomond and the Trossachs

One focus of our exploratory analysis was to identify recurrent mentions of places across different guide books and authors. GATE allows us to quickly identify the number of place names from Loch Lomond and the Trossachs in each text document. Because many place names are repeatedly used in describing a certain geographic extent, we can also extract the place names themselves in the order in which they are described. This allows us to analyse which routes the travel writers were following. It is noteworthy that most guides follow a pre-described route that traces previous literary work of the area, most notably that of Sir Walter Scott and his poem 'The Lady of the Lake', which popularised the landscapes of Loch Lomond and the Trossachs for tourism:

> '[...] there are scenes so grand, so magnificent, and so exquisitely beautiful, that it is a matter of surprise they lay unnoticed and comparatively unknown in the midst of our land, like buried gems, till near the beginning of the present century, when Sir Walter Scott's matchless poem "The Lady of the Lake," flashed across the length and breadth of Britain, and invested the Trosachs with an interest which, we are persuaded, shall never die away.'
> (Nelson's Tourist Guide, 1858)

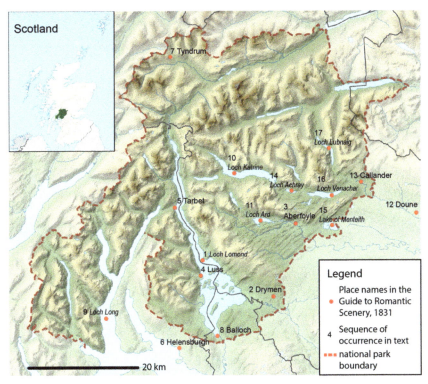

Figure 7.3: Places within the study area mentioned in the 'Guide to Romantic Scenery' numbered according to first occurrence. (base map credit: Wikimedia user Nilfanion, map created using Ordnance Survey open data CC-BY-SA-3.0).

For further analysis, we mapped the identified place names of our study area in a Geographic Information System and compared these maps between the different writers (Figures 7.3 & 7.4). This follows a methodology that has been previously applied in the field of Digital Humanities to highlight spatial differences between the journeys of different writers, for instance, in the English Lake District (Donaldson, Gregory, and Taylor, 2017a; Gregory and Donaldson, 2016). For our approach, we extracted for each of our documents the place names that were matched in our gazetteer of Loch Lomond and The Trossachs place names. In case these were mentioned several times in a document, we chose to use the first occurrence of each place name, and numbered the place names according to the sequence in which they occurred in each document. We then assigned coordinates to each place name using the GeoNames gazetteer[3] with a focus on the UK. We manually disambiguated terms where two or more

---

[3] http://www.geonames.org/

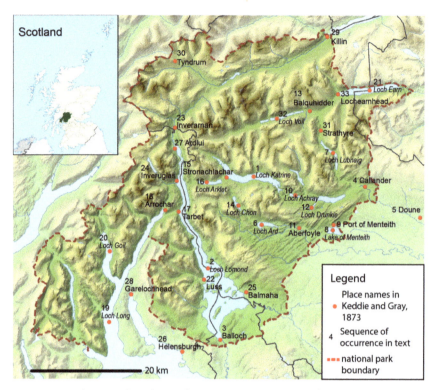

Figure 7.4: Places within the study area mentioned in the 'Keddie and Gray' numbered according to first occurrence. (base map credit: Wikimedia user Nilfanion, map created using Ordnance Survey open data CC-BY-SA-3.0).

matches were found within the UK (e.g., for Balloch we chose the settlement in West Dunbartonshire and not Balloch as a populated place in the Highlands). For place names where there was no match in GeoNames, we manually added the coordinates that were returned by Google Maps. In the open source Geographic Information System QGIS (QGIS Development Team, 2018), we displayed the coordinates on an OpenStreetMap base map and labelled them in the order in which they occurred.

Taking the map of place names from the Guide to the Romantic Scenery (Figure 7.3) and comparing it with the Guide by Keddie and Gray, we observe that fewer place names from our list occurred in the Guide to Romantic Scenery, and that they focused on the South and South East of what today is the Loch Lomond and The Trossachs. In Keddie and Gray, the place names found also include the north-eastern part of the study area around Loch Earn (Figure 7.4).

Looking at the sequences of place names, there often did not seem to be an understandable pattern (Figure 7.5). This contradicts our finding from our more close-reading work, where we observed similar trajectories being

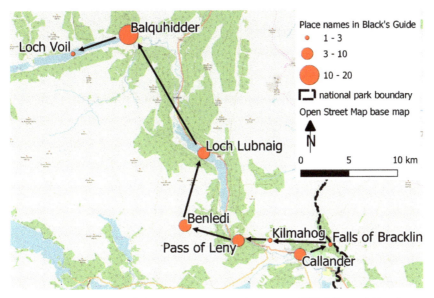

Figure 7.5: Sequence of places names in a subsection of 'Keddie and Gray' scaled according to the number of mentions. Basemap © OpenStreetMap contributors.

described. Going back to the original text sources this is explained by the fact that many guidebooks provide a summary of the travels at the beginning, such as in Keddy and Gray:

> *Starting from Glasgow, the tourist has the choice of proceeding by railway to **Balloch**, thence sailing up **Loch Lomond** to **Inversnaid**, then crossing the country by coach to **Loch Katrine**, proceeding thence to **Callander**, where the train may be taken either for Glasgow or Edinburgh.*

As we selected the first occurrence of each place name, we thus mapped the introduction section that mentioned a low granularity 'broad-brush' geographic description. Or, as in 'Nelson's', the section on Loch Lomond starts with a mention of historical figures such as Rob Roy and places that were important in their lives, which does contain places that were necessarily visited by the writers. For a more detailed analysis, it would therefore be important to distinguish between places mentioned in the guidebooks and places visited by the writers (Cooper and Gregory, 2011). In order to better map the travels of the writers in a spatial sense, we chose to focus on a subsection of one of the guidebooks for illustrative purposes, also manually annotating additional place names which we had missed by using a simple and limited gazetteer.

This second case study focused on the sections in Black's Guidebook that described the geography around Callander and Balquidder. We manually searched for all the locations in the document and annotated their first occurrence, as well as the number of occurrences thereafter for places that were visited. After mapping this itinerary and scaling the symbol for each place name according to the frequency of mentions in this section of the document, we identified three additional place names that were not contained in our gazetteer: *Falls of Bracklin, Pass of Leny and Benledi* (a hill), which we also then manually georeferenced using Google Maps.

According to this map, which has a finer granularity of the text due to its focus on a smaller section within a guide book, we are able to observe a finer spatial granularity of the travels described, which map out the route of Black's description of the Trossachs.

In addition to comparing the places and the succession in which they were visited by travel writers, it also proves fruitful to analyse how the same landscapes were described by different writers. For some of the recurrently mentioned touristic places, such as Loch Achray (Figure 7.6), we found that the descriptions were similar in the praises of the scenic experience the landscape offered. Loch Achray (Figure 7.3) is described in The Steam- boat companion; and Stranger's Guide to the Western islands and Highlands of Scotland (1820):

> "*Loch Achray, which is very romantic, being closely wooded from the brink of the water, to the top of the almost perpendicular hills that surround it.*"

Figure 7.6: Contemporary view of Loch Achray (photograph by F. Wartmann).

Alternatively, Nelson's tourist guide (1858) describes Loch Achray as:

"[···] the glassy waters may be seen gleaming through the leafy curtain like a sparkling gem. On a calm evening the lovely scene wears an aspect of deep seclusion and tranquillity."

However, we also found some cases of differences in descriptions. For instance, the transition between the Highlands and Lowlands around the town of Callander was described as *'the cold swelling moorlands which connect the Lowlands with the Highlands'* and the *'dreariness of the scenery is apt to be acutely felt'* in Black's Guide (1853). Callander itself was described as *'a mongrel sort of village, neither Highland nor lowland — some of the dirt and laziness it has of the former, and some of the hard stony and slaty comfort of the latter'* in Black's Guide (1853), but as *'worth a visit'* in The Steam-boat Companion (1820) and as *'beautifully situated upon the banks of the river Teith, immediately upon the confines of the Highlands, and surrounded with woods and scenery of the most romantic description'* in Lumsden's 'Guide to the romantic scenery of Loch-Lomond, Loch-Ketturin, the Trosachs' (1831). Such differences in the description of similar geographic areas are interesting because they offer glimpses into how different writers have perceived the landscape. On the one hand, such descriptions can then be compared between different writers within a certain time period (e.g., the Romantic period). On the other hand, these descriptions may also be taken together from different writers within a time period and compared to writings of a different time period in order to trace changes in landscape perception that may be linked with actual physical landscape changes.

7.2.4   *Application example 2: Words about landscape – Analysing historic landscape descriptions through term co-occurrence*

Another aspect worth highlighting is the possibility to use a text mining tool such as GATE to automatically locate mentions of particular terms that are used to describe landscape experience. For instance, our close-reading highlighted that landscape terms (e.g., *mountains, lake, scenery*) often seemed to be co-occurring with descriptors such as *picturesque, romantic, terrible* or *sublime*, as is exemplified in this passage from Chapman:

'Whatever, indeed, is beautiful, or fantastic, or wild, or picturesque, or sublime, or terrible, are associated in this celebrated region.' (The picture of Glasgow, and strangers' guide; with a sketch of a tour to Loch-Lomond, R Chapman [1818] p. 314).

We therefore conducted a more quantitative analysis of the co-occurrences between our list of landscape terms and these landscape descriptors across our entire corpus. Using the ANNIC tool in GATE which enables complex searches of patterns in annotated text, a list of all the landscape terms found in the corpus was compiled, based on the output of the original corpus annotation. We also manually compiled a list comprising relevant descriptors, for which we chose: *beautiful, picturesque, romantic, sublime, tranquil* and *terrible*. In GATE, we converted both lists into gazetteers and developed a set of simple JAPE rules (Cunningham, Maynard, and Tablan, 2000) to automatically annotate sentences which contained at least one example of both a descriptor and a landscape term. We developed an application comprising the gazetteer lookup and rules, and ran it over the annotated corpus. Using ANNIC, every occurrence of such a sentence in the corpus was extracted and exported to a spreadsheet. Finally, we used a Python script to find all co-occurrences of landscape and descriptor term within these extracted sentences, and output a co-occurrence pair for each, along with the document in which it was found. We then aggregated the documents according to the book from which they were derived (e.g., Shearer's Guide).

This analysis showed that the terms *beautiful* and *picturesque* co-occurred most often with landscape terms, but with some differences between authors (Table 7.2). For example, when normalising the number of co-occurrences with the total number of tokens to compare between authors, Shearer's Guide and Nelson's Guide are highlighted as containing higher ratios of co-occurrences with the term *beautiful* than Chapman's, Black's or Keddie and Gray's guidebooks. Furthermore, The Guide to Romantic Scenery by Lumsden (1831) contains more co-occurrences of landscape terms with *picturesque* than with *beautiful* (Figure 7.7). The term *sublime* is much less used by these authors, and comparatively more so by Lumsden (both 1820 and 1831) and Chapman (1895). Differences between the use of landscape descriptors in historic travel writings was also reported for a corpus on the Lake District, where Donaldson, Gregory, and Taylor (2017) reported differences in the frequency of use of the descriptors *beautiful, picturesque, sublime* and *majestic* between works from the Romantic period and the Victorian era.

Examples from our corpus indicate that an analysis of such terms always has to take into account context, as we otherwise risk misinterpretations. For instance, the terms may be negated (e.g., 'anything but picturesque') or be used in a manner where the descriptor does not describe the landscape terms, as exemplified in Black's guide to the Trossachs, 1853 describing the village (landscape term) of St. Fillans:

> 'It is altogether a show pet village, with its allotments and trellises of creeping flowers, and more adapted to the philanthropist than the searcher after the **sublime** and **terrible**.'

In using the measure of co-occurrence, we are making an assumption that if a term such as *beautiful* or *picturesque* occurs together in a sentence with a

Black Shilling		Steam Boat Companion		Edinburgh and Glasgow		Guide to Romantic Scenery		Nelsons		Picture of Glasgow		Shearer's	
scenery	1	scenery	10	lake	3	castle	1	stream	1	scenery	5	ruin	2
mountain	1	mountain	2	scenery	3	lake	1	view	1	castle	3	valley	2
		channel	1	mountain	2	landscape	1			village	2	castle	1
		loch	1	ruin	2	road	1			ascent	1	house	1
		mill	1	archipelago	1	rocks	1			mill	1	scenery	1
		rock	1	buildings	1	scenery	1			mountains	1		
		ruin	1	hills	1	village	1			river	1		
				rock	1					rock	1		
				town	1					site	1		
				woods	1					trees	1		
										valley	1		

Table 7.2: *picturesque:* Number of co-occurrences for landscape terms with the descriptor.

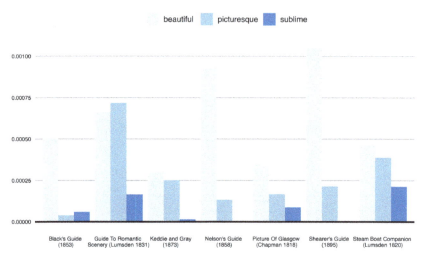

Figure 7.7: Frequencies of co-occurrences for *beautiful, picturesque* and *sublime* with landscape terms.

landscape term, then they are in some way related through the context of the sentence, such as '*Many **beautiful glens** are on the estate*' (Shearer's). Here, the terms *beautiful* and *glen* are grammatically and contextually related. However, in more complex sentences this is not necessarily the case, as in: '*the **beautiful** residence of Sir Patrick Murray, near which are the **falls** of the Turret, with the **glen** and **loch** of the same name*' (Picture of Glasgow), where *beautiful* refers to *residence*, and is not a descriptor of the landscape terms *falls, glen* or *loch*. Furthermore, this case also illustrates that we may find several landscape terms co-occurring with a single descriptor such as *beautiful*. In that case, it is unclear to which landscape term the descriptor refers to. To resolve this, full linguistic parsing would need to be added to the system in order to obtain grammatical relations. However, this introduces problems as it is not so reliable, especially on these kinds of texts. We therefore also conducted a manual analysis of the identified co-occurrences (e.g., Table 7.2), with two goals in mind. First, we wanted to determine in how many cases the co-occurrences of descriptor and landscape term reflected the semantic context of the sentence. Second, we used it to identify the landscape term(s) to which the descriptors referred.

Of a total of 326 co-occurrences, 264 were identified as reflecting a descriptor referring to a landscape term, indicating that co-occurrence within a sentence is a useful, albeit rough, approximation. Most co-occurrences referred to the scenic appreciation of landscape elements such as the lochs, mountains and rivers of Loch Lomond. In 39 cases, the landscape descriptor referred to 'scenery' and in 33 cases to 'view' or 'views'.

In contrast to scenic visual experiences, we found few mentions of acoustic experience. For example, the term *tranquil* only co-occurs with landscape terms nine times in our corpus (four times are in Keddie and Gray and three times in Chapman 1818). Although the term *tranquil* encompasses both acoustic and visual experience, it serves as an indicator that visual experience dominates in touristic landscape descriptions of Loch Lomond. This is in contrast to work on a historic corpus of Lake District writing, where particular soundscapes, as well as quietness and tranquillity appear to have been more central to the experience of the landscape (Chesnokova et al., 2019; Taylor, 2018).

Through manual inspection of the documents in our corpus, we did find some examples for acoustic descriptions of rivers or streams, e.g.:

> 'It [the Keltic stream] does not flow through a valley, but tumbles, wild and brawling and speckled with foam, down a narrow rocky gorge, roaring and leaping as it goes, in a succession of dashing falls [⋯]' (Nelson's Tourist Guide, 1858),

or:

> 'The roaring of the mountain torrents in a calm morning after a raining night was something quite delicious to my ears, and actually makes a kind of music, of which you dwellers in the plains can have no conception. From the platform before our door, we had twenty at least in sight, and more than a hundred within hearing; and the sort of thrilling they made in the air, with the mingling of the different waters, on the least swelling of the breeze, had an effect quite overpowering and sublime'. (Keddie and Gray, 1873).

Several guides also describe specific locations where an echo can be produced at Loch Katrine and Loch Ard. The fascination with echoes by travellers especially in the late 18th and early 19th centuries has been described in detail for the Lake District, where travellers were fascinated by the potential of the landscape to generate powerful echoes (Taylor, 2018). However, while we did find some instances of acoustic landscape appreciation, the results of the co-occurrence analysis indicate a dominance of visual appreciation of landscape over other senses our corpus. Using the co-occurrence of landscape terms and their descriptors can thus provide insights into how landscape is described in terms of auditory and visual information, and how the respective importance of these aspects may shift over time.

7.2.5 Challenges in applying natural language processing to an analysis of historic travel writing

Because we only used a small set of place names, we also manually checked the text documents to identify additional texts with place names from Loch

Lomond that were not yet contained in our lists. During this process of manual annotation of additional place names, we observed that the historic corpus contained many spelling variations, which we did not pick up automatically using the contemporary spelling of Loch Lomond place names (indicated in bold in the list below). These historic spelling variations should therefore be included in our place names list, for instance:

- Loch-Ketturrin / Loch Ketturin /Loch Catrine / Loch Katerine / **Loch Katrine**
- Loch- Venu-Char/ Lake Venachoir/ Loch Vennachar / **Loch Venachar**;
- Loch-Lomond / **Loch Lomond**;
- Callender / **Callander**;
- Trosachs / **Trossachs**;
- Loch-Archlat / **Loch Arklet**;
- Arroquhar / **Arrochar**

Even though we missed many spelling variations, a number of Loch Lomond place names were identified in the texts, which enabled us to quickly zoom into the geographically relevant text passages.

Using the annotations with our place names and landscape terms in combination with a close reading also highlights the need to expand the lists of landscape terms specifically to our corpus, which is illustrated with the following text example (blue: identified place names, green: identified landscape terms, pink: place names not identified, yellow: relevant landscape terms not contained in the list we used for searching):

'From this point to Loch Katrine *the* glen *receives the name of the* Trosachs *. It is flanked on the right by the lofty* summit *of* Ben-an *; and* Ben Yenue *[sic] (2388 feet) rears its stately* crest *on the left. [...]*

Every turn of the road *unfolds fresh* views *of* wild *and romantic* beauty *, on which the eye reposes with new delight. The* valley *is one continued maze of rugged* mountains *, grey* rocks *, and green* woods *, lofty* precipices *and dark* ravines *, shivering* cliffs *and heathery knolls, with masses of* trees *dispersed in* picturesque *confusion, and, conspicuous amongst the sylvan beauties of the* landscape *, the light and graceful* Highland birch *, which, singly or in tufts, clambers up the tall* precipices *, and adapts itself, as no other* tree *has the power of doing, to the varying features of the fairy-like* scene *. Amidst all this amplitude and diversity of form, the eye is surprised by the ever-shifting effects of light and shade, producing all the day long a succession of novel and striking pictures'.*

(in Edinburgh & Glasgow to Stirling: Doune, Callander, Lake of Menteith, Loch Ard, Loch Achray, the Trosachs, Loch Katrine, Loch Lomond, Keddie and Gray [1873], p. 35).

Particularly, we would need to include terminology used in Victorian travel writing and landscape descriptions that include words such as 'sublime', 'picturesque', 'terrible', 'views', 'scene', 'precipices', which are used to describe landscape, which we identified as relevant during our analysis, but were not contained in our original gazetteer of landscape terms. Furthermore, we observed that some texts that had been processed using OCR contained many errors, which made it impossible for the gazetteer lookup to identify the landscape terms and place names (Table 7.3). It is possible to match in a more flexible way, or even to first normalise spellings based on methods such as edit distance, as is sometimes used in social media processing (Maynard, Bontcheva, and Augenstein, 2016), but these methods can also over-generate and should be used with caution.

These challenges notwithstanding, this exploratory analysis of Victorian travel writing aided by automated text processing shows some of the possibilities that text processing toolkits offer to transition between distant reading and close reading. It should be borne in mind that while automated language processing tools offer the possibility to process large amounts of text very quickly, thereby saving laborious human effort, nevertheless results are almost never 100% accurate, so caution should always be taken.

7.3   Reading the Trails as an Environmental Historian

One of the exercises we felt offered a useful test of the GATE analysis tool was to compare its findings with a traditional assessment of the material compiled by a close historical reading. By conducting this parallel 'forensic fishing' activity,

Optical Character Recognition Text	Original Text
Looking to the ndrth, a frightful preciffice df the mountain is seen, 2000 feet deep; while stretching «ftr as the eye Can l^eadb, is seen the tremendous assemblage of ruggedness that constitutes the Gramt- l^an chain.	*Looking to the north, a frightful precipice of the mountain is seen, 2000 feet deep; while stretching as far the eye can reach, is seen the tremendous assemblage of ruggedness that constitutes the Gram-pian chain.*
On the bold diib of BenTenne	*on the bold cliffs of Benvenue*

Table 7.3: Example of digitised and original text in the Guide to the romantic scenery of Loch-Lomond, Loch-Ketturin, the Trosachs, James Lumsden, (1831).

we were looking to ascertain two things: firstly, how the data from a GATE analysis might embellish, challenge and/or inform a conventional methodological/archival trawl using environmental humanities techniques, and, secondly, what might be 'missing' from a strict linguistic analysis in terms of historical context or 'buried' information hard to decode from the data alone. What follows is a short summary – from an environmental historian's perspective—of the historical value of the corpus, alongside a few reflections on how different processes of data collection and interpretation might be usefully aligned.

Firstly, it seems helpful to flag the fact that issues of place-making, landscape change and the relationship between material and imagined spaces have long occupied the attentions of environmental historians. More specifically, the construction of particular kinds of landscapes as places of unique natural value (especially in terms of how those ideal types fed historical conservation movements) has been a particular focus of scholarship, manifested, for instance, in writing on the creation of national parks. Equally, a comprehensive historiography exists on the politics of landscape and shifting attitudes towards the 'wild', aesthetics and Romanticism; the complex relationship between nature tourism, technology and modernity; and issues of empire and environmental colonialism. As such, this is the broader methodological context by which the environmental historian approaches the physical and textual landscape of Loch Lomond and the selected materials under study.

Deconstruction of the corpus of guidebooks using the techniques associated with a close historical reading ably illuminates the emergence of Loch Lomond as a visitor destination in the 1800s and highlights the importance of gazetteer literature in fostering and curating that tourist experience. *The Steamboat Companion; and Stranger's Guide to the Western islands and Highlands of Scotland* (1820) made direct reference to the role of its genre in raising consciousness as to the glories of a trip to the Highlands. As this pamphlet noted, before the publication of Thomas Pennant's *Tour of Scotland* (1769) – the first guidebook to the region – Scotland was viewed as a 'land of barbarism and misrule', a forbidding frontier landscape that might be compared to 'wild Africa'. By the time of *The Steam-boat Companion*, however, it had been recast as a terrain of 'unusual grandeur', terrifically rather than terrifyingly wild and 'capable of fascinating the most indifferent observer, and rousing his profound admiration'.

Tracing an empirical trail using our corpus of guidebooks, we can sequentially plot the historical construction of Scotland as a rugged paradise and, in the process, interrogate the findings of the GATE data analysis above. Read as a collective, these precursory *Lonely Planet* guides provide a ready inventory of historical detail as to what the aspiring tourist should see and how to get there. As such, the textual path set out in these guides gives a clear sense of the emergence of a set of key landmarks that each traveller had to 'tick off' on their expeditions. The repetition of these places in successive texts (as well as the recitation of their distinct qualities) established a core list of 'must-do' visits that communicated a clear landscape perception of the region's scenic quali-ties. Elevated in these guides, then, are an inventory of aesthetic treasures, Loch Lomond (and Ben Lomond), Lochs Ness and Katrine, the Falls of Clyde, the Trossachs, Skye and Glencoe, all of which were installed as iconic spaces of nature tourism. Performing a conventional historical reading alongside GATE analysis allows for a valuable charting, comparison and mapping of this emerg-ing literary geography.

The Loch Lomond vacation experience was not entirely a case of 'roughing it' however. Embedded in the guidebooks is a sense of a gathering infrastructure of tourist provision – information on inns, roads, steamships and hotels to ease the visitor's journey. This 'hidden' context would be hard to extract from data mining alone. Also implicit in the historical record is a cogent route map with a defined choreography to it, a directed movement from one site to the next. *Nelson's Tourist Guide* (1858) pointed out that the pathways of travel were of major import, that one should operate 'upon the principle that it is more pleasing to be led from soft and tame to magnificent and sublime than the reverse'. Aesthetics dictated the trails to be followed: another subtlety laid bare by a qualitative reading. *The picture of Glasgow, and strangers' guide* (1818) by R. Chapman advised the guidebook should be a necessary 'companion' to take on any excursion.

Where the GATE analysis is especially useful is in providing a data set to consider alongside a historical textual-crunch of guidebook landscape descrip-tors. The guides certainly spent a lot of time detailing the scenic delights of the Scottish Highlands as a place of wildness, antiquity and unique quality. Writers commonly referenced its rugged and mysterious qualities, and especially its keenly felt connection with the Romantic ideal and untended views that inspired contemplation, awe and a dose of terror. Descriptions abounded of glorious lakes and mountains, short offerings on flora and fauna, geology and hydrology, all wrapped up in a textual veneer of wild charm. The view from the top of Ben Lomond was frequently situated as a supreme site of wilderness pilgrimage, to *Shearer's Guide to Stirling, Dunblane, Callender, the Trossachs and Loch Lomond* (1895), a 'noble panorama'. *Nelson's Tourist Guide* (1858) reminded tourists that the climb really was not that hard. Common to these descriptions was a sense of remoteness, emptiness and solitude. What GATE adds to a historical reading of such passages is the opportunity to formally catalogue, delving (as we did above) into the text to see the relative use of terms such as 'beautiful', 'picturesque' or 'sublime'. Sometimes a traditional reading of guidebook content mitigates against this through a preference for sweeping prose. As Chapman commented, Loch Lomond and its environs possessed '**almost every requisite** [emphasis added] to exalt the imagination, to engage the attention, to sweeten peace, and to furnish amusement to solitude'.

As successive guidebooks built on one and another to construct a Highlands nature resort for well-to-do English visitors, scattered notations showed the other contributing factors abetting the process. Royal patronage was one factor – Queen Victoria having first visited in 1842 and Balmoral purchased in 1852. *Shearer's Guide* reprinted an account of a royal visit in 1869 in full. More significant (and, indeed, a factor in the Queen's attachment) was the raft of literary figures writing rapturously of Highlands landscape, in particular Sir Walter Scott and his famous 1810 *Lady of the Lake* epic poem (referenced and reproduced in many guides including *Shearer's* and *Nelsons* and which sold 25,000 copies in its first six months of publication). The importance of royal or literary patrons created a sense of 'value' to the nature collections on display, but not one easily quantified. What is, however, more easily traceable is the gathering repetition of phrases associated with Scott, for example, or of oft-repeated phrases attributed to Loch Lomond that conferred on it an identity as a lake famed for its 'fish without fins, waves without wind, and a floating island' (referenced in *Edinburgh & Glasgow to Stirling: Doune, Callander, Lake of Menteith, Loch Ard, Loch Achray, the Trosachs, Loch Katrine, Loch Lomond, Keddie and Gray* [1873]).

Contained in the guides are frequent comparisons, either to the pastoral lines of the English countryside; or to 'blue-chip' nature sites elsewhere in the world. A common theme in both was the supposition that Scotland (and particularly the Highlands) denoted a space apart, an unmediated slice of primeval rugged land that had escaped the transformations of industrialism. *A Journey to the Western Isles of Scotland* (1775) by Samuel Johnson, pointed out that this 'unknown and untravelled wilderness' inspired sensations very different from those of the 'artificial solitude of [English] parks and gardens' in nurturing a haunted sense of 'want and misery, and danger'. Here, a historical reading can provide useful contextual reference points in thinking through the parallel histories of designed landscapes elsewhere. In *The Guide to the romantic scenery of Loch-Lomond, Loch-Ketturin, the Trosachs* (1831) James Lumsden encapsulated an important relationship between scenic landscapes and cultural nationalism in ranking Loch Lomond as the best lake in Great Britain and superior to Lakes Geneva and Constance for 'wild magnificence and solemn grandeur'. Joined to the heroic mantle of raw nature was a sense of ancientness – antiquity commonly depicted in the form of old castles and cathedrals. Lumsden noted that visitors scanned the scenery with 'anxious eye' searching 'for the haunts of those whom history has chronicled'. Here the environmental historian can make a valuable intervention in exploring the tenuous threads connecting landscape, people and constructions of identity. Embedded in the text, too, was a definite sense of the imperial gaze – often expressed in the presentation of Highlander culture as warlike, lawless and primitive. Samuel Johnson talked of communities in possession of a 'savage wildness' and (importantly) made reference to the political dynamics behind the creation of a tourist landscape in the form of the Highland Clearances.

7.4   Concluding Thoughts

To the environmental historian, these guidebooks show a tourist landscape under construction, one that crafted a consistent narrative of celebrating the wild and the romantic, and communicated that via a recurring phraseology and a clear choreography of the visitor experience based on visual markers. Text analysis with GATE throws up new and provocative possibilities in analysing this material by allowing us to identify sections of texts with relevant landscape descriptions; to map a literary landscape more forensically and to compare different authors over time, for example, by comparing enumerations of different terminologies used; and by quantitatively cataloguing the use of terms such as 'beautiful' and 'picturesque', which have frequently been discussed in landscape history.

There remain a few caveats to this approach. The first of these involves the issue of concealed messages buried deep in the text, as well as the value of illustrations, appendices and supplementary material which an automated text mining approach fails to capture. By the time of the *Shearer's Guide*, a raft of companies catered to the every whim of visiting tourists – in the appendix, a wealth of adverts can be found touting steamboats, outfitters, bookshops and hotels and 'hydropathic spas' – and illuminated the networks of industrialism beneath the lakes and mountains. These glimpses are important, hints as to the hidden stories behind the text and the importance of reading *absence* as well as presence in the narrative tracks of elite tourism in the Scottish Highlands. Significant here is what a strictly defined linguistic analysis does not communicate: namely the broad context of political processes of land seizure as well as the voices of historical actors – Highlanders, women, local tour guides – who scarcely appeared in the scripted trails of gazetteer literature.

That said, by highlighting the possibilities of 'forensic fishing' in practice, this chapter usefully illustrates the way in which text processing and historical methods can work together to construct an innovative and blended approach to environmental questions of landscape and place-making. Answering such questions is relevant for landscape research that seeks to look at landscape not as a snapshot in time, but taking into account a historical perspective of landscapes as dynamic and ever-changing. Landscape research seeks to understand people–place relations, and while much of the focus is on contemporary relations of people and places, integrating historical information on how people perceived, interacted with and described landscapes of the past helps to improve our understanding of how these relations have changed and develop to the present, and to shed light into how they may be evolving in the future.

Furthermore, studying historic relations between people and landscapes is not only relevant from a landscape research perspective, but may also inform contemporary policy-making. For example, policies that seek to restore landscape to some historical point in time (such as certain rewilding or landscape conservation initiatives) may be challenged by notions of landscape found in these historic guidebooks that highlight the historically rooted and manifold relations between people and places that they visited and inhabited.

References

Baird, D N L (1853). *Baird's Guide through the City of Glasgow, and to Loch Lomond, the Falls of Clyde.* Glasgow, p. 31.

Brydone, J. and sons (1856). *Brydone's Guide to the Trosachs, Loch Lomond, the Highlands of Perthshire, etc.* Ed. by J. Brydone. Edinburgh, p. 88.

Chesnokova, Olga, Joanna E. Taylor, Ian N. Gregory, and Ross S. Purves (2019). "Hearing the silence: Finding the middle ground in the spatial humanities? Extracting and comparing perceived silence and tranquillity in the English Lake District". In: *International Journal of Geographical Information Science* 33.12, pp. 2430–2454. ISSN: 1365-8816. DOI: 10 . 1080 / 13658816 . 2018 . 1552789.

Cooper, David and Ian N. Gregory (2011). "Mapping the english lake district: A literary GIS". In: *Transactions of the Institute of British Geographers* 36, pp. 89–108. ISSN: 00202754. DOI: 10.1111/j.1475-5661.2010.00405.x.

Cunningham, Hamish, Diana Maynard, and Kalina Bontcheva (2011). *Text Processing with GATE.* California: Gateway Press. ISBN: 0956599311.

Cunningham, Hamish, Diana Maynard, and Valentin Tablan (Dec. 2000). "JAPE: A Java Annotation Patterns Engine". In:

Donaldson, Christopher E, Ian N Gregory, and Joanna E Taylor (2017a). "Implementing corpus analysis and GIS to examine historical accounts of the English Lake District". English. In: *Historical atlas.* Ed. by Peter Bol. Northeast Asian History Foundation, pp. 152–172. ISBN: 9788961874564.

— (2017b). "Locating the beautiful, picturesque, sublime and majestic: Spatially analysing the application of aesthetic terminology in descriptions of the English Lake District". In: *Journal of Historical Geography* 56, pp. 43–60. ISSN: 03057488. DOI: 10.1016/j.jhg.2017.01.006.

Gregory, Ian and Christopher Donaldson (2016). "Geographical text analysis: Digital cartographies of Lake District literature". English. In: *Literary mapping in the digital age.* Ed. by David Cooper, Chris Donaldson, and Patricia Murrieta-Flores. Digital Research in the Arts and Humanities. Routledge, pp. 85–105. ISBN: 9781472441300. DOI: 10.4324/9781315592596-14.

Keddy, W. and Gray (1873). *Edinburgh & Glasgow to Stirling, Doune, Callander, Lake of Menteith, Loch Ard, Loch Achray, the Trosachs, Loch Katrine, Loch Lomond.* Glasgow: Maclure and Macdonald.

Maynard, Diana, Kalina Bontcheva, and Isabelle Augenstein (Dec. 2016). *Natural language processing for the semantic web.* Vol. 6, pp. 1–194. DOI: 10.2200/ S00741ED1V01Y201611WBE015.

Purves, Ross S., Alistair J. Edwardes, and Jo Wood (2011). "Describing place through user generated content". In: *First Monday. Peer-Reviewed Journal on the Internet* 16.9. DOI: 10.5210/fm.v16i9.3710.

QGIS Development Team (2018). *QGIS Geographic Information System.* Beaverton: Open Source Geospatial Foundation. URL: http://qgis.osgeo.org.

Richardson, T. (1798). *Guide to Loch Lomond, Loch Long, Loch Fine, and Inverary, With a Description of All the Towns, Villages, etc.* Glasgow: R. Chapman.

Shaw, G. (1878). *Shaw's Tourist's Picturesque Guide to the Trosachs, Loch Lomond, Central Highlands.* London.

Taylor, Joanna E. (2018). "Echoes in the mountains: The romantic lake district's soundscape". In: *Studies in Romanticism* 57.3, pp. 383–406. ISSN: 00393762. DOI: 10.1353/srm.2018.0022.

CHAPTER 8

The Wild Process: Constructing Multi-Scalar Environmental Narratives

Joanna E. Taylor

Department of English, American Studies, and Creative Writing, University of Manchester, UK

Benjamin Adams

Department of Computer Science and Software Engineering, University of Canterbury, Christchurch, New Zealand

In *The Song of the Earth*, arguably the foundational text of 21st-century ecocriticism, Jonathan Bate writes that the ecopoet is someone who can turn the experiences of 'reverie, solitude, [and] walking' into language (Bate, 2001, p. 42). His approach to reading is based on a similar methodology: through specific close readings of carefully chosen poems, Bate demonstrates how poets such as William Wordsworth (1770–1850) (re-)wrote the British environmental narrative to prioritise individual responses to a natural world. In doing so, each poet — though in very different ways — sought to situate themselves as part of an ecosystem in which the same forces of life and joy 'rolled through all things' (Wordsworth, "Tintern Abbey", in *Major Works* (Wordsworth, 2011)).

Bate's argument tends to assume that all people have the same kind of access to green spaces, and so can share similar ecopoetic experiences which – by

How to cite this book chapter:
Taylor, Joanna E. and Benjamin Adams (2022). "The Wild Process: Constructing Multi-Scalar Environmental Narratives." In: *Unlocking Environmental Narratives: Towards Understanding Human Environment Interactions through Computational Text Analysis*. Ed. by Ross S. Purves, Olga Koblet, and Benjamin Adams. London: Ubiquity Press, pp. 161–178. DOI: https://doi.org/10.5334/bcs.h. License: CC-BY 4.0

extension – can be unpacked and understood through an ecopoetic analytic. But the experiences that poets like William Wordsworth express, and that Bate explores, are almost exclusively predicated on the assumption that the ecopoet goes forth into the landscape alone, ready to enjoy a solitary communion with 'Mother Nature'. Bate sees no problem in assuming that masculine pronouns can stand in for all human experience, and that nature can be figured as female; it's a practice 'as old as Hesiod', he writes. But Bate's unwillingness to challenge this ancient power dynamic between an implicitly male ecopoet and a feminised landscape has wide-reaching consequences that, as we explore here, extend beyond the close readings of historical environmental narratives.

Scott Hess has pointed out that there are distinct problems with understanding the experiences of authors such as William Wordsworth as being representative (Hess, 2012, pp. 10–11). The problem, as Jos Smith indicates in *The New Nature Writing*, is that it is impossible to effect the distinction between nature and culture which this kind of ecopoetics assumes: 'Beyond the aesthetic production of "Nature" as a particular style of distant and enshrined object', he asks, 'how might those engaged with the world do things differently?' (Smith, 2017, p. 15). This concern is born out of a tradition Smith shares with Bate: his worry that landscape is becoming 'distant' is a direct consequence of the individualism – at worst, even close-mindedness – of Bate's Wordsworthian ecopoetics.

But none of these critics offer substantial alternatives to this model, which Kathleen Jamie has called that of the 'Lone Enraptured Male' (Jamie, 2008). For Jamie, the consequences go beyond the literary: they are at the very heart of what we mean when we talk about 'wildness' in landscapes – and, even more than that, are core to the practices we develop to manage 'wild' places. What we offer here is a challenge to the individualistic, ecopoetic approach to British environmental narratives by using Jamie's thinking as inspiration for both the texts we use and the methods we employ to analyse them. Following Jamie's guidance to incorporate and acknowledge more – and more diverse – voices into environmental narratives, and to situate community at the heart of how we interpret both text and landscape, what we describe here is an attempt to unite distance and closeness, individualism and community, and computational analysis with human reading. In short, what we want to suggest is that, by attending to alternative voices in environmental narratives, we can find new methodological perspectives, too.

In this instance, we have turned to two corpora of environmental writing by British women to investigate, first, how women's environmental narratives differ from men's and, second, what implications for multiscalar approaches to text and landscape we can uncover from these overlooked works. The first is a small corpus of writing about Rannoch Moor that extends the collection introduced earlier in this volume, and allows us to consider how masculine forms of 'wildness' differ from feminine ones at a specific location whose very geography stimulates considerations on the ways we negotiate closeness and distance. We then situate the readings of this corpus as part of an analysis of a larger corpus of nature writing that allows us to investigate at scale the implications of gender on environmental narratives. First, though, we turn to Jamie to develop a more thorough understanding of why gender matters for the ways we read, interpret and deploy environmental narratives.

8.1   Challenging the 'Lone Enraptured Male'

In her review for the *London Review of Books* of Robert Macfarlane's *The Wild Places* (Jamie, 2008), Jamie challenges Macfarlane's understanding of what it means to express the experience of being in a wild place. Macfarlane's celebration of the 'Lone Enraptured Male' represents a figure who imagines that he is '[h]ere to boldly go, 'discovering', then quelling our harsh and lovely and sometimes difficult land with his civilised lyrical words' (Jamie, 2008). In his writing, and in that of the authors in whose footsteps Macfarlane imagines himself to be treading (not least William Wordsworth), the landscape is made bare of people; Macfarlane deliberately seeks a place that feels like it is 'outside of human histories'. Instead, it becomes a place characterised by 'silence, an avoidance of voices other than the author's, just wind in the trees, or waves, the cry of the curlew' (Jamie, 2008), or what Jamie later calls 'theatrically empty places' in which emptiness and wildness might be performed, but is never genuine. In this tradition, Jamie argues, human histories – particular those of the peoples who have worked the landscape – are overwritten in favour of a carefully crafted version of nature which tells us much more about the individual author than the environment they describe. For Jamie, Macfarlane's 'lovely honeyed prose' lays 'an enchantment on the land': nature is thus something Other, something to be tamed by the 'enraptured' literary response of the 'bright, healthy and highly educated young man', who packages the landscape for the armchair geographer's willing consumption (Jamie, 2008).

The issue Jamie identifies here is far higher stakes than only the literary. Accounts like Macfarlane's indicate a form of environmental management; in this case, an individual's narrative is imposed on the wild place, and that account in turn comes to influence the way the land itself is managed. This is perhaps most obvious in the English Lake District, which was designated a UNESCO World Heritage Site in 2017 on the promise that it would be managed as a 'cultural landscape' on the strength of its long-standing literary, artistic, and agricultural histories (Lake District National Park Partnership, 2016). Jamie believes that, when nature and ecopoetic culture come together in this way, ecodiversity itself is threatened:

> When the wild is protected by management, or re-created by the removal of traces of human history, you have to ask, who are these

managers? Why do conservationists favour this species over that? Whose traces are considered worth saving, whose fit only to be bull-dozed? (Jamie, 2008)

The problem with narratives that seek to elide human histories in favour of a mythic version of 'wildness' is not only that they risk stripping the landscape of its cultural diversity: they also risk promoting versions of environmental man-agement that homogenise a wild place's flora and fauna. A lack of diversity in literary representation, in Jamie's approach, matters because it leads to a break-down of ecodiversity as a result of prioritising one kind of individualist narrative over the multitudinous alternatives.

But, Jamie argues, there is another approach to both reading and writing environmental narratives which offers new ways of understanding the nature writing genre and what we mean by 'wildness'. Rather than emphasising – as both historical and contemporary accounts of nature writing have tended to — the 'lone enraptured' response to a place, Jamie advocates for amplifying the voices of those myriads of people, 'many of them women', who study, think and write about the environment as a complex ecosystem in which nature and culture interlink. Macfarlane also recognises this problem with his intuitive approach: towards the end of *The Wild Places*, he realises that it is 'nonsensi-cal' and 'improper' to think of anywhere as being 'outside of human histories'. Jamie notes appreciatively that Macfarlane's recognition of his 'myopia', which has caused him to look 'too much into the apparently empty distance', leads to a revelation about the multiscalar nature of wild places: 'a wild place is not nec-essarily landscape-sized, and not necessarily an adventure playground. A wild place can also be mouse or beetle-landscape sized, and everywhere, and near at hand' (Jamie, 2008). Reading the environment through a multiscalar lens, Jamie implies, brings the distance close and imposes a sense of enormity onto the tiny. It is only when we apply this approach to both reading and writing envi-ronmental narratives that we can uncover the diversities – eco- and human – which make up any wild place.

This argument for diversity – wherein class, race and gender are also brought to bear on the environmental narrative – encourages Jamie to reform the notion of wildness. It is not a place, she concludes, but might instead be 'bet-ter described as a process'. When wildness becomes 'a force requiring constant negotiation' (Jamie, 2008), it is able to contain a huge variety of plants, animals, cultures and voices that respond to changing cultural and climatological envi-ronments. Refusing to pin down a wild place's environmental narrative allows for more creative and more adaptable responses, whether those take the form of literary or management decisions. How this diversification operates at a liter-ary level is our focus for the remainder of this chapter. Specifically, we want to ask what impact gender has on the narrating of environmental landscapes and experiences. To do so, we follow other contributors to this volume by starting at a particular location that has both a literarily and environmentally diverse history: Rannoch Moor.

8.2   Processing Literary Diversity on Rannoch Moor

Our starting point was two pieces of nature writing: extracts about the Moor from W. H. Murray's account *Undiscovered Scotland* (Murray, 2003), and Robert Macfarlane's *The Wild Places* (Macfarlane, 2008). Because our interest in this case was specifically in nature writing as a literary genre, we took these extracts and added four more to them to create a small corpus of 19th-, 20th- and 21st-century nature writing about Rannoch Moor. Our adapted corpus included the extracts from Murray and Macfarlane, alongside a passage from Dorothy Wordsworth's (William's sister) *Recollections of a Tour Made in Scotland* (Wordsworth, 1997) (first written in 1803), Kathleen Jamie's poem 'The Way We Live' (Jamie, 1987), an extract from Linda Cracknell's 2014 account of her pedestrian adventures *Doubling Back* (Cracknell, 2014), and Jackie Kay's poem "Rannoch Loop" (Kay, 2017), written about a reading in Rannoch's remote train station as part of her role as Scotland's Makar.

This small corpus totalled just over 6700 words, and there are of course problems with using a corpus of this size to extract meaningful patterns from textual data. But bringing Bate's literary ecopoetic approach into conversation with the distant readings promoted by a computational one that favour quantitative or quantifiable data allows us to generate multiscalar modes of textual analysis that offer new opportunities to create diverse environmental narratives out of a corpus of texts, of the kind for which Jamie advocates. In other words, when we use these works as part of a multiscalar process, the goal – to echo Jamie – is not to pin these works down to a particular reading, but to constitute them as part of the 'complex negotiation' of ongoing creations of meaning that slide between distance and closeness.

The starting point in this process, as with Jamie's mode of viewing wild places, is to start with a distant view that encompasses the whole landscape. In literary scholarship – as in environmental studies – gaps or rarities are often as meaningful as what is present, and in the case of distant reading that encourages an approach that emphasises key patterns, either of data that is noticeably present or noticeably absent. To begin this process, we loaded the raw text into AntConc, a free package for basic quantitative text analysis (Anthony, 2004). In this case, the opening question was simple enough: Is Jamie right that women's nature writing does something different to the masculine tradition with which we are more familiar? Jamie herself helpfully suggests an entry point for this kind of enquiry: she finds that, in the style of nature writing cultivated by Macfarlane and other in his tradition, there is 'an awful lot of "I"' (Jamie, 2008). A straightforward query, then, is whether or not Macfarlane and Murray use 'I' more often than Wordsworth, Jamie, Cracknell and Kay. What we would expect to find, if Jamie's deliberately polemical view is right, is that the female authors use terms that indicate community more frequently than writers from the individualistic masculine tradition.

Using the collocation tool, we can ascertain which words are statistically most likely to appear in connection with each other. Since what we are interested in is the written representation of a particular place, it makes sense to use 'Rannoch' as the search term. The resultant lists (Table 8.1) indicate which words are most closely associated with Rannoch in the texts by men and those by women.

Collocate	t-score
'Rannoch' Men (Murray, Macfarlane)	
Moor	3.12
Miles	2.62
Loch	2.4
Fifteen	1.73
Out	1.72
My	1.67
I	1.53
Yards	1.4
Distance	1.4
Rannoch	1.36
'Rannoch' Women (Wordsworth, Jamie, Cracknell, Kay)	
Moor	2.23
Rest	2
Best	2
You	1.73
Refreshed	1.41
Finally	1.41
Dear	1.41
Boat	1.41
Beloved	1.41
Walking	1.4
Loch	1.4
Lake	1.4
Rannoch	1.4
Your	1

Table 8.1: Collocated words with 'Rannoch' by organised authors' genders.

Here, we have excluded prepositions from the most significant 25 collocations in order to foreground nouns that indicate the things about the Moor that these authors find significant.

These collocations indicate that Jamie's proposition – that the 'Lone Enraptured Male' is particularly enamoured of his solitary experiences – is true; the significance of personal pronouns ('I' and 'my') to both Macfarlane and Murray suggests that both authors focus on their personal responses to the landscape. In these cases, that relationship is figured as an ecopoetic subject/object split, where nature is viewed as a static thing to be traversed by the lone wanderer – and this notwithstanding the fact that neither Murray nor Macfarlane travel alone (Macfarlane is with his father, and Murray with his dog). Indeed, these authors' careful documentation of distances ('miles', 'yards') indicates how important the quantification of the landscape is in this tradition – even if, as Macfarlane explains, for the slow-going across the peaty moor, distances should be 'measured in hours, not miles'. Partly, this desire to measure indicates the profound discomfort that arises from the fact that the Moor resists this kind of quantification: Macfarlane finds that its 'vastness and self-similarity' – peat hags and boggy fissures repeating themselves for miles without much to distinguish them – makes it impossible to judge distances by any standard measurements. It seems to him that they make no progress across the landscape: 'like explorers walking against the spin of pack ice, our feet fell exactly where we had lifted them'. Only the movement of the prose registers that they are, in fact, moving forward, despite what their perception of the geography suggests.

The collocates for the four texts by women are very different; in fact, the only overlaps are 'moor', 'loch' and 'Rannoch'. The remainder of the collocates indicate an alternative mode of constructing an environmental narrative. The most significant collocates in the women's works can be roughly divided into two main categories: connection and embodiment. In these texts, Rannoch Moor is not the site of 'perfect solitude' that Murray describes; rather, it is an isolated spot that inspires profound feelings of interaction between the author and peoples both past and present, as well as with the Moor itself. The significance of 'you' and 'your' (and absence of the 'I' and 'my' which characterises the men's collocates) indicates the extent to which these authors find at the Moor a site for conversation: Jamie addresses an unnamed 'you'; Cracknell begins her walk with friends, and then turns her attention to the absent peoples whose traces are everywhere evident on the Moor; and Kay addresses the memory of her father, who walked here in his own hiking days. Wordsworth's use of 'your' is more complicated; she refers to her brother and walking companion, as well as the local cottagers with whom they stay. But she also directly addresses her reader, Mary Wordsworth, to whom she was writing along the tour: 'you will not wonder that we longed to rest', Wordsworth tells her after a particularly long day. Wordsworth's personal 'you' contrasts with Macfarlane's generic reader: 'while you sleep', he advises, the sleeper train 'conjur[es] you to a different land'. Other collocates ('dear', 'beloved') further

she arrives on the moor in the late afternoon, she finds a densely populated land-scape: 'pylons stalking the Fort William railway line; an occasional Scots Pine isolating itself as a dark silhouette, flattened by dull light. I followed a quad bike trail to find the "creep", a low gap under the embanked railway'. In Cracknell's reading, the railway line is not romantic, but evidence of the Moor's modern technological entwinement with the rest of the country. By contrast, the con-temporary trees – unlike Macfarlane's ancient forests – are alone, anomalies in the flat light of the expansive moorland. The quad bike trails, meanwhile, are evidence of communities nearby, and of the fact that the Moor – for all its apparent isolation – is a worked landscape. Each of these details confirms that Rannoch Moor is core to the interlinking of complicated communities, of which it is, by implication, a part.

As she carries on walking, Cracknell also finds that the peat bogs force her into a different mode of movement. But Cracknell, unlike Macfarlane, does not try and overcome the terrain: she moves with it:

> Alone, the meshing of rhythm, thought and observation had me invent-ing songs and rhymes. Lyrics were delivered in my head to the tune of Walking on the Moon by The Police.
> 'I hope your legs don't break
> Walking Rannoch Moor.
> A boat's what you should take
> Walking Rannoch Moor'.

The scenery inspires her towards very specific forms of writing: lyrics which 'mesh' the rhythms of her walking with those of the landscape and her lan-guage, and which collaborate with both the environment and songs she knows. This is not a poem designed to rest silently on the page – it is designed to be sung, tested out, in a way that creates a kind of vicarious collaboration between moorland, reader and writer (and, by extension, even The Police). Even as she walks alone across Rannoch Moor, Cracknell discovers a communal experience that generates a collaborative environmental narrative.

There is, then, a striking difference between the wildness of Macfarlane's Rannoch, and the wildness of Cracknell's. Cracknell's narrative falls into line with how Jamie thinks wildness should be understood: somewhere 'theatrically empty [...] peopled by ghosts' and, crucially, somewhere – or something – that requires 'constant negotiation'. Macfarlane, meanwhile, exaggerates the Moor's isolation, even as he crosses it in company. These individuals are representative of this small corpus, then, in displaying a distinctively gendered approach to interpreting this landscape.

It is hopefully evident that even a small corpus of eight documents can lead us towards new interpretations of environmental narratives. However, such a small corpus does mean that the results are not generalisable, and we might come to very different conclusions if we replaced these documents with an alternative set. A central question remains: Are environmental narratives distinguishable by gender at scale? By expanding the type of a multiscalar approach we have described so far, we can begin to ascertain whether these patterns repeat at a larger scale across other kinds of environmental writing.

8.4   The Country Diary

In order to compile a larger corpus of historical nature writing, we turned to the 'Country Diary' column in the *Guardian newspaper*, a liberal British publication that has been in circulation since 1821. Originally a regional newspaper titled the *Manchester Guardian*, in 1959 it changed its name to reflect its growing reputation as a national broadsheet. 'The Country Diary' column began in 1906, when the paper was still predominantly regional, and has remained a feature throughout the newspaper's evolution. Combining natural history, environmental reporting and reflections on the natural world at local and national scales, the column represents a genre that differs markedly in both aims and form from the creative non-fiction on which we focused in the smaller corpus. Whilst the texts we have looked at so far are interested in imaginative, social and cultural interpretations of the environment, 'The Country Diary' is explicitly interested in natural history and rural issues; it is a good example of the *Guardian*'s mission to acknowledge that 'comment is free, but facts are sacred'[1]. Unlike the texts we have looked at so far, then, its aim is not necessarily to unpack individuals' private responses to a particular place. Nevertheless, taken together the entries in this column offer cogent examples through which to explore patterns in how male and female authors have written environmental narratives.

Digital raw text copies of over 6000 Country Diary articles are available for download on the Guardian Open Platform[2], along with associated metadata (including the author's name, and the byline for each article). We used the application programming interface (API) for the Open Platform to download these articles, and then we utilised the Genderize API to assign a gender to each article. The Genderize API provided a likely gender based on the forename in the byline, along with an associated probability value. The gender of the vast majority of authors could be identified with a high probability. In the small number of cases where a gender could not be identified, it was because either 1) the forename was only an initial, 2) the forename was one that is used as both a male and female name (e.g., Carey), 3) there were multiple authors, including both men and women, or 4) it was an anonymous author (e.g., editor or letters). We removed these unidentifiable authors from our dataset, leaving us with a set

---

[1] C. P. Scott, 'A Hundred Years', (May 1921)
   https://www.theguardian.com/sustainability/cp-scott-centenary-essay
[2] https://open-platform.theguardian.com

Category	Count
All articles	6600
Articles with gender identified	6174
Articles by female authors	1385
Articles by male authors	4789
Unique female authors	66
Unique male authors	80

Table 8.2: Counts of articles in the Country Diary corpus by category.

of 6174 (out of the original 6600) articles. Overall, the column has historically been written by a reasonably small group (80) of men. Despite the fact that, in terms of unique authors, 45.2% are women versus 54.8% men (Table 8.2), the Genderize API suggested that 22.4% were authored by a woman and 77.6% were authored by men. The metadata provided by the open access dataset does not allow us to assess change over time, but that could of course be a factor.

Having identified the authors' genders, the first test we attempted was to see if it is possible to train a general classifier on this corpus based on the two author-ship categories (male and female). The goal in doing this was to determine if a simple model can differentiate between male and female authorship, based on an author's choice of diction. We used a Naïve Bayes classifier implemented in the Mallet toolkit[3] because it is a relatively simple model that is quick to train. First, we created a balanced dataset containing all 1365 articles by women and a random selection of an equal number of articles by men. The summary results for the Naïve Bayes classifier, using a held-out set of 10% of the data for test-ing and 10-fold cross validation, are shown in Table 8.3. The high values for accuracy, precision, recall and f1 all indicate that the classifier can identify two categories of documents which each use distinctive terms. In other words, this sample seems to support our readings of the Rannoch Moor corpus to suggest that men and women do construct environmental narratives differently.

What is less clear, however, is whether these differences are always attributable entirely to gender or whether they represent the practices of indi-vidual authors. Because so many articles are written by the same authors, it is possible that the classifier is capturing something specific about the language used by the small number of authors who have written several Country Diary articles, about the particular practices of the column's editors (the last four of whom have been women), or about the specific genre of writing that the col-umn represents. Specific answers to these queries are beyond our scope here,

---

[3] http://mallet.cs.umass.edu/classification.php

Balanced set of 2770 articles	Mean	Std. dev.
Train accuracy	0.995	0.001
Test accuracy	0.866	0.030
Test precision (female)	0.892	0.032
Test precision (male)	0.845	0.040
Test recall (female)	0.833	0.049
Test recall (male)	0.899	0.034
Test f1 (female)	0.861	0.032
Test f1 (male)	0.871	0.029

Table 8.3: Naïve Bayes results on a balanced set of 2770 articles.

Balanced set of 120 unique authors	Mean	Std. dev.
Train accuracy	0.995	0.005
Test accuracy	0.525	0.112
Test precision (female)	0.514	0.129
Test precision (male)	0.541	0.179
Test recall (female)	0.527	0.201
Test recall (male)	0.527	0.111
Test f1 (female)	0.505	0.146
Test f1 (male)	0.518	0.130

Table 8.4: Naïve Bayes results on a balanced set of 120 unique authors.

but what we can do is to examine the individual biases that the column's authors bring to the dataset. To explore that, we created a much smaller and more balanced corpus of 120 articles, 60 by unique female authors and 60 by unique male authors. As we noted with regards to the Rannoch Moor corpus, computational approaches like this text classifier do a poorer job of differentiating the gender categories with less information. The results, as shown in Table 8.4, indicate that it is essentially random whether an article is classified as being written by a male or female author. The high *train accuracy* versus the low *test accuracy* also indicates that the method is perhaps *overfitting* the data. This is not to say that there is not a difference in the language that these authors use, but in this case our method does not provide any further insight into that.

By combining this approach with the attention to individual authors' language, as we did with the Rannoch Moor corpus, however, we can explore in more detail the nuances behind these general trends. We again used the

Author category	'I' per article	'we' per article
Female (2770 articles)	2.61	1.23
Male (2770 articles)	2.83	1.38
Female (unique authors)	4.83	1.54
Male (unique authors)	3.73	1.56

Table 8.5: Counts of personal pronouns per article.

AntConc software to look at the use and context of the personal pronouns 'I' and 'we' in the two balanced corpora. The data for counts 'I' and 'we' are shown in Table 8.5. The first corpus (balanced set of 2770 articles) shows that female authors use fewer personal pronouns (both 'I' and 'we') than male authors, but the difference is rather slight. In the smaller corpus of 160 unique authors, female authors actually use 'I' at a higher rate than male authors (4.83 vs. 3.73). In addition, for both categories of authors the rate is much higher than in the larger corpus. Partly, this consistent high rate is thanks to the nature of the genre; the newspaper column is dependent on personal accounts that lean towards the factual.

However, as we saw in the Rannoch Moor texts, the contexts in which these pronouns are deployed indicate a complexity behind their use for which numbers alone do not account. These hidden nuances are indicated in the terms with which 'we' clusters most commonly across the two corpora. For male authors, 'we' clusters most frequently with 'were', 'had', 'could', 'have', and 'are'. For female authors, the top four terms are similar: 'have', 'are', 'were' and 'had' (the fifth, 'walked', anticipates the further differences which we turn to below). A subtle yet important distinction begins to emerge in this comparison, because across the corpus present-tense prepositions are more likely to indicate general stances (e.g., 'we are a nation of wildlife lovers'; 'we know dolphins eat fish and we are comfortable with it') as well as specific experiences (e.g., 'we are almost stunned with our good luck'; 'soon we are driving past the first houses in Llandegfan'). The order in which these prepositions cluster with 'we' matters, then, because it indicates that women are more likely to merge their specific experiences (described by the writing 'I') with their readers' imagined feelings. To put it another way, in the male authors' articles, 'we' describes a circumscribed set of people to whom the action described in the article has happened. For the female authors, on the other hand, 'we' is more likely to indicate an imagined community that includes the reader alongside the writer.

Comparably subtle, but nevertheless important, differences emerge when we perform a collocate analysis of the two terms (Table 8.6). Similarly to what we did in the earlier study, prepositions are excluded. In addition, we have not included different tenses and plurals for terms more than once. For the unique author corpus the minimum collocate frequency was set at four, and for the

Pronoun	Author category	Collocates
I	Female (2770 articles)	realise, remember, suspect, crouch, surprised, myself, shall, think, noticed, read, aware, notice, know, hear, pleased, feel, lucky, discover
	Female (unique authors)	ashamed, realised, wouldn't, suspect, surprised, speak, hadn't, faires, anything, remember, feel, pause, hear, wanted, finding, wish, reached
	Male (2770 articles)	confess, realised, wish, wondered, suspect, suppose, remember, myself, recall, think, knew, noticed, peer, hear, listen, counted, love
	Male (unique authors)	suspect, remember, wondered, realised, waited, believe, think, wish, walked, alarm, watched, myself, bike, spot, aware, saw, wanted
we	Female (2770 articles)	descend, shall, walked, watched, met, know, pass, visited, saw, went, heard, hope, looked, returned, leave, cross, later
	Female (unique authors)	damage, pass, reach, knew, hope, need, making, stone, follow, thought, school, come, walk, always, know, said
	Male (2770 articles)	ourselves, approached, went, climbed, crossed, walked, passed, reached, met, saw, watched, found, got, know, stood, our, upon
	Male (unique authors)	decide, eat, hope, went, review, reached, hear, know, sure, do, want, walked, should, saw, anything

Table 8.6: Collocated words for the pronouns 'I' and 'we' in the different corpora.

larger corpus it was set at 10. This allows us to focus on terms that show up in a number of different articles. The window size for collocation in both cases was set at five words to the left and right. The t-scores are not shown for brevity but all range from around five to seven.

We can see here that female authors tend to employ an affective, and affecting, vocabulary that more vividly captures their emotional responses to the landscape: they are ashamed or pleased, feel lucky or surprised and express longing ('want' and 'wish') for absent things. Male authors, meanwhile, are more likely to focus on the head than the heart: they remember or realise, wonder and think. And these differences are echoed in the way men and women seem to look at the environments about which they write: male authors are more likely to observe a place from a distance (they peer, spot, watch and notice), whilst female authors are more likely to focus on hearing, a sense that both relies on a more circumscribed geography and which tends to be a more serendipitous experience rather than something sought out. That might be because women move more slowly; they are more likely to 'crouch' and 'pause', allowing more time for discoveries and, crucially, for building relationships with the flora and fauna that make up the local environment. One author neatly compares the effects of pausing and speed on a grey wagtail she encounters at the edge of the A27:

> During the past few weeks this individual has become increasingly confiding. When I pause just a few feet away from where it is feeding, it continues trotting down the path towards me, its long black and white tail pumping furiously, and head flicking from side to side as it sets its sights on a swarm of midges that is rising up to wreath my head. A cyclist swooshes into the tunnel, ringing his bell. The wagtail takes flight, uttering a single metallic 'tchik' call as it flits between the paddles of the replica water wheel that sits in the preserved remains of the old mill race.

Her slowness, both in this moment and in her patience over the 'past few weeks', is rewarded by the wagtail's trust; its happy 'trotting down the path' indicates a sense of bonhomie between author and bird. The moment is disrupted by a male cyclist (it is notable that 'bike' is one of the key collocates for solitary male authors), who rushes past in a way that disrupts both the calm and the quiet shared by the wagtail and the column's writer – and the wagtail's 'tchik' implies that it feels a similar irritation to the author at this disruption. Passages like this might make us wonder what the male authors – at least, those intent on walking, cycling, climbing, crossing or passing – are missing when they do not take the time to develop this kind of slow connection with their environment. And it pays off: these encounters, facilitated by patience, are reiterated by the sense of 'luck' and 'discovery' that women are also more likely to experience throughout this corpus.

There is, undeniably, significant overlap between the genders in this dataset, and in how they narrate environmental narratives. These similarities are an inevitable result of the close rules of this genre – a very specific newspaper column with a particular, quasi-scientific agenda – but they perhaps make the subtle differences even more telling. This corpus reiterates the divergences between male and female authors that we saw in the Rannoch Moor corpus in ways that, at first glance, might seem incidental. Taking the time to close read particular texts, though, has a similar effect to pausing in the pursuit of a 'Country Diary': pausing allows us to notice how meaningful small differences can be.

8.5   Conclusion

The multiscalar combination of close and distant reading techniques we have outlined here emphasises the extent to which human experiences of the natural world are always heterogeneous. Nevertheless, we can generalise about some of these differences based on an author's gender. The fall-out of this gendered approach to both environment and the narratives it inspires has, as Jamie believes, serious consequences for how we narrate – both in the sense of writing about and managing – the green spaces with which we live today. These differences might be taken as opportunities: in the outliers, we might uncover new directions for environmental narrative and, from there, action. As we have tried to show here, the same process is necessary for both reading texts and managing environments. Both practical and literary narratives depend upon multiscalar negotiations between distance and closeness, and the willingness of the reader and policy-maker to unpack the connections and interactions that operate between them.

References

Anthony, Laurence (2004). "AntConc: A learner and classroom friendly, multi-platform corpus analysis toolkit". In: *Proceedings of IWLeL*, pp. 7–13.

Bate, Jonathan (2001). *The Song of the Earth*. London: Pan Macmillan.

Cracknell, Linda (2014). *Doubling back: Ten paths trodden in memory*. Glasgow: Freight Books.

Hess, Scott (2012). *William Wordsworth and the ecology of authorship: The roots of environmentalism in nineteenth-century culture*. Charlottesville: University of Virginia Press.

Jamie, Kathleen (1987). *The way we live*. Hexham: Bloodaxe Books.

— (2008). "A lone enraptured male". In: *London Review of Books* 6, pp. 25–27.

Kay, Jackie (2017). *Bantam*. London: Picador.

Lake District National Park Partnership (2016). *Nomination of the English Lake District for inscription on the World Heritage List (2015-16)*. https://www.lakedistrict.gov.uk/caringfor/projects/whs/lake-district-nomination. [accessed 19 October 2016].

Macfarlane, Robert (2008). *The wild places*. London: Granta Books.

Murray, W.H. (2003). *Undiscovered Scotland*. Macclesfield: Baton Wicks Publications, pp. 155–163.

Smith, Jos (2017). *The new nature writing: Rethinking the literature of place*. London: Bloomsbury Publishing. DOI: 10.5040/9781474275040.

Wordsworth, Dorothy (1997). *Recollections of a tour made in Scotland*. London: Yale University Press. DOI: 10.2307/j.ctt1ww3v9z.

Wordsworth, William (2011). *William Wordsworth: The major works*. Ed. by Stephen Gill. Oxford: Oxford University Press, USA.

CHAPTER 9

Inferring Value: A Multiscalar Analysis of Landscape Character Assessments

Joanna E. Taylor
Department of English, American Studies, and Creative Writing, University of Manchester, UK

Meladel Mistica
Melbourne Data Analytics Platform, University of Melbourne, Australia

Graham Fairclough
McCord Centre for Landscape, Newcastle University, UK

Timothy Baldwin
Department of Natural Language Processing, The Mohamed bin Zayed University of Artificial Intelligence & School of Computing and Information Systems, The University of Melbourne, Australia

What is considered environmentally 'valuable' varies enormously from place to place, and from time to time, but some common denominators can be identified. A landscape assessment for the Wanganui District[1] of the north island of New Zealand, for example, puts the emphasis on memory and affect:

---

[1] Wanganui District: Outstanding Natural Landscape Assessment, 10th July 2015. https://www.whanganui.govt.nz/files/assets/public/district-plan-changes/hudson-landscape-assessment-introduction-10-7-2015.pdf

How to cite this book chapter:
Taylor, Joanna E., Meladel Mistica, Graham Fairclough, and Timothy Baldwin (2022). "Inferring Value: A Multiscalar Analysis of Landscape Character Assessments." In: *Unlocking Environmental Narratives: Towards Understanding Human Environment Interactions through Computational Text Analysis.* Ed. by Ross S. Purves, Olga Koblet, and Benjamin Adams. London: Ubiquity Press, pp. 179–196. DOI: https://doi.org/10.5334/bcs.i. License: CC-BY 4.0

'A landscape becomes memorable when the image perceived by the viewer remains with them after they leave the site'. The capacity of a place, landscape or site to influence a visitor emotionally – and the durability of that affect – leaves the most 'valuable' legacy, both in economic and personal terms: it is the private memory of a place which encourages repeat visits, and concern for the future survival of a location's particular character.

There are long-standing variations in how 'value' is applied nationally and locally (Bloomer Tweedale Architects and Town Planners, 1992). The Wanganui assessment continues, 'it is not possible to fully define what makes landscapes memorable, as the combination of factors is numerous and of different importance to different people'. In the UK Government's National Planning Policy Framework 2019 'value' is closely aligned with significance, which is assessed according to the 'value of a heritage asset to this and future generations because of its heritage interest' (Ministry of Housing, Communities and Local Government, 2019). That 'interest' might be archaeological, architectural, artistic or historic in nature (Ministry of Housing, Communities and Local Government, 2019, p. 71), and elsewhere recreational and environmental values are also included (Ministry of Housing, Communities and Local Government, 2019, p. 29). However, as planning advisors CSA Environmental observe, the NPPF 'does not define a "valued landscape"'; instead, what 'value' means is left open to judgement (CSA Environmental, 2017). Indeed, one of the goals of a baseline study may be to define what 'value' is taken to mean for a particular practical purpose or in a certain place (Landscape Institute, 2013, p. 32). Combining computational and human-led text analysis, our goal is to determine some common factors in these diverse definitions.

The multiplicity of definitions of landscape value has encouraged legal disputes over whether or not, or how far, a landscape is environmentally, socially, or culturally valuable. For example, in England in 2015 one case pitted Stroud District Council against the Department for Communities and Local Government, forcing a clearer definition of landscape value in that particular context, notably that a 'designation' – for instance, as an Area of Outstanding Natural Beauty – was not a prerequisite for 'value'. However, it also clarified that landscape value should be based on 'demonstrable physical attributes rather than just popularity'.

Formal assessments of landscape value can take place at many scales, from national to local. Nationally recognised areas with high landscape value are formalised into statutory designations with varying levels of protections in the UK, for example, as National Parks and Areas of Outstanding Natural Beauty. Local authorities at district or county level may also recognise special areas, using nomenclature such as Areas of Great Landscape Value (AGLV) or similar terms[2]. An AGLV acknowledges a place's emotional significance to the

---

[2] https://www.planningportal.co.uk/directory_record/347/local_landscape_designation_for_example_area_of_high_landscape_value

local population, as well as its cultural, environmental or economic value. In Scotland, these local designations have been amalgamated into one category: National Scenic Areas, which aim to represent the variety of features that might be considered truly 'Scottish': prominent landforms, coastline, lochs, rivers, woodlands and mountains (Office, 1996). These places tend to lack the consistency of an AONB or National Park; their value is more closely aligned with local feeling than objectified scenery (Institute, 1996).

Since about 1990, these assignations of value have been identified and described in Landscape Character Assessments (LCAs). As Graham Fairclough explains in Chapter 2, an LCA seeks 'to be holistic in its understanding of landscape', but often tends to emphasise the present, physical state of a place over historic and affective markers. As Fairclough also observes, LCAs are further complicated by their vocabulary choices. Calling an area a 'landscape' indicates a close relationship between people and place. A landscape's identity is predicated on the interplay between human residents and natural features, and as such value is accorded to those locations that display this kind of relationship over a sustained period of time (Landscape Institute, 2013, p. 14). A landscape, then, does not necessarily need any special attributes; indeed, its everyday usage – its apparent ordinariness – might be the very thing that makes it valuable.

It should be clear, then, that assessing 'value' depends on complex and iterative interplay between national legislation and local contexts, between expert and lay assumptions, between a range of academic and professional practices and between local residents and perhaps more distant but nevertheless deeply engaged interests. Nor can LCA-type exercises ever be totally separate from political, ideological and economic factors; LCAs always take place in a particular cultural and social context. Landscape characterisation, in other words, is a concept that relies on multiscalar negotiations. In this chapter, we take inspiration from the construction of the environmental narratives represented by a corpus of LCAs to develop a multiscalar methodology that situates local understandings of landscape value in national contexts. In doing so, our aim is to uncover some of the political and cultural assumptions that underpin notions of value in British Landscape Character Assessments, and to draw out the challenges of applying these human assumptions to environmental management and landscape planning.

9.1 Methodology

9.1.1 Creating a corpus

The origins and history of LCA (and its 'cousin' Historic Landscape Characterisation [HLC]) since the late 1980s were set out by Carys Swanwick and Graham Fairclough as part of an overview of the idea of landscape characterisation as exercised in Britain, Europe and further afield (Fairclough, Sarlöv Herlin, and

Swanwick, 2018). Critically, LCA was conceived in the later 1980s in opposition to the 'traditional' approach to countryside protection. That approach, exemplified in the British National Parks and Access to the Countryside Act of 1949, was based on formal designation of sharply defined blocks of land singled out from the seamless whole of the 'entire territory'. These were areas that for one reason or another were considered more special, or 'outstanding', or nationally important. For most of the 20th century, these assessments were based on matters such as visual connoisseurship, an appreciation of remoteness and the romantic allure of largely non-industrialized and rustic environments. In contrast, whilst still seeking to create a practical tool for landscape protection (or rather, particularly as the method evolved, landscape management), LCA aimed to define and promote the recognition of landscape character everywhere, whether 'good' or 'bad', valued or in need of improvement and change. LCA, in short, aimed to move away from a post-Romantic understanding of landscape to a wide-reaching recognition of the ways that people and place intersect.

LCA sought to be as interdisciplinary and holistic as possible, recognising the need to include perspectives on landscape other than only the visual or the affective: ecological, historical and so on. It was adopted as vehicle by a range of professionals and practices following specific goals, whether those of landscape archaeologists and heritage managers seeking a broader view of the past-in-the-present, to nature conservationists rebranding themselves as biodiversity champions and recently transforming into ecosystem analysts, to social scientists seeing landscape as a forum for 'capturing' public perceptions and aspirations for the landscape in which they lived and so on. A typical LCA, if there is such a thing, will probably have been largely written by geographers or landscape architects, but will contain several other voices, vocabularies and visions, and frequently an ecological or environmental sentiment will be dominant.

In these assessments, the definition of 'character areas' allows a heterogenous combination of character traits to be connected to an area without any need to give it a value relative to other areas, whether adjacent or distant. For land management purposes, it is sufficient to be able to discuss in each area what gives that particular area its distinctive identity, as defined in terms of expert knowledge or of lay, local or affective interest. To say that these character areas are not assigned a 'value' does not, of course, mean that various values and different nodes of valuation are not embedded in the identification of their characteristics or the definition of their boundaries. LCAs are not totally objective, but they aim to defer absolute statements of significance and importance — one dominant form of 'value' in landscape management, protected area designation and spatial planning — until the point of need, when proposals and projects are being discussed that might adversely affect the character of a landscape. The methods used by LCA can be replicated by other people to produce different results, depending, for example, on which aspects of landscape are given priority, or indeed how certain dimensions or aspects are interpreted: approaches to the concept of 'nature' vary widely across disciplines, for instance.

This innate heterogeneity, and the fact that LCAs have now been produced at various scales for over 30 years, creates a great diversity, and choosing examples for our corpus was not straightforward. We initially intended to compare LCAs (or their equivalents) from several countries in order to try to assess national cultural difference to how landscape is perceived, but the complexities of dealing with multiple languages was deferred to a later period of research. Thus restricted to LCAs written in English, we chose examples of LCAs from England, Wales and Scotland. We examined LCAs carried out at various scales, and both free-standing and as part of overarching and coordinated national surveys: some county-scale LCAs in England[3] and a selection of the 159 area descriptions from the England 'National Character Areas' (NCA) assessment[4] (formerly in the 1990s in its original iteration named the 'Countryside Character Map') currently curated on the web by the government agency Natural England.[5] The latter NCA descriptions were largely selected to bring in areas that are not amongst those popularly or traditionally viewed as special landscapes (i.e., not national parks or Areas of Outstanding Natural Beauty (AONBs)). Two of the county scale LCAs (Derbyshire and Cumbria), on the other hand, include the Peak and Lake District National Parks. Beyond England within the UK,

---

[3] County scale England LCAs: Derbyshire https://www.derbyshire.gov.uk/ environment/conservation/landscapecharacter/landscape-character.aspx; parts of the Devon LCA [North Devon] https://www.northdevon.gov.uk/media/290514/ north-devon-torridge-lca-191110.pdf, East Sussex: https://www.eastsussex.gov.uk/ environment/landscape/ North Norfolk: https://www.north-norfolk.gov.uk/media/ 1271/landscape_character_assessment.pdf, and the Lake District https://www.lakedistrict.gov.uk/data/assets/pdf_file/0020/170480/landscape_ character_assessment.pdf

[4] https://www.gov.uk/government/publications/national-character-area-profiles- data-for-local-decision-making/national-character-area-profiles#other-sources- of-information

[5] From the national England LCAS: NCA15, Durham Magnesian Limestone Plateau: http://publications.naturalengland.org.uk/file/8461491; NCA68, Needwood & South Derbyshire claylands: http://publications.naturalengland.org.uk/ file/4472935; NCA86, South Suffolk / North Essex claylands, http://publications. naturalengland.org.uk/file/5148978341478400; NCA115 Thames Valley, http://publications.naturalengland.org.uk/file/6085686941712384; NCA149, The Culm (Devon), http://publications.naturalengland.org.uk/file/5462962095521792.

we examined 24 of the 48 area descriptions in the all-Wales 'National Landscape Character' assessment,[6] and some Scottish equivalents.

This selection provided a variety of different topographies and land-uses allowing us to explore ways that different landscape types led LCA in different directions. We also selected areas which have followed different trajectories of protection, change and urbanisation/commercialisation through the 20th century, most notably by including some LCAs covering protected areas such as National Parks where relatively little 'modern' development has changed older landscapes, and other LCAs concerned with areas where change has been much less constrained: thus the LCA treatment of both special areas and so-called 'ordinary, everyday' landscapes can be compared. A further level of diversity was introduced by the different disciplinary backgrounds and interests of the teams producing each LCA, although the national LCA descriptions subsequently went through a process of standardisation. Finally, though we focus only on British examples, the constituent countries have distinct cultural attitudes to landscape.

9.1.2   Parsing the corpus

Although there are detailed guidelines on what is included in an LCA, each individual council has the freedom to interpret these guidelines according to their needs. There is also diversity in the balance between different aspects of landscape (some privilege natural heritage, others focus on historical information, others on the scenic and the visual; some attempt all). The structure and display of LCA documents also varies. Whilst some LCA documents are unembellished texts, others are eye-catching publications with precisely placed images, decorated borders and colour motifs, foregrounding certain aspects and guiding the reader and user (e.g., a decision-taker in environmental planning) towards particular 'significant' findings. Early LCAs in the 1990s and even later were printed, usually informally for local use although occasionally more formally published. Nowadays, LCAs tend to be uploaded to official websites for download as Portable Document Formats (PDFs), often in several PDFs divided by geographical area or by landscape theme. We have limited ourselves to LCAs available online in electronic form, and therefore to the more recent examples.

---

[6]   Wales: nlca02-central-anglesey; nlca04-llyn; nlca06-snowdonia-description; nlca08-north-wales-coast-; nlca10-denbigh-moors; nlca12-clwydian-range; nlca14-maelor; nlca16-y-berwyn; nlca18-shropshire-hills-outliers; nlca20-radnorshire-hills; nlca22-aberdyfi-and-coast; nlca24-ceredigion-coast; nlca26-upper-wye-valley; nlca28-epynt; nlca30-brecon-beacons-and-black-mountains; nlca32-wye-valley-and-wentwood; nlca34-gwent-levels; nlca36-vale-of-glamorgan; nlca38-swansea-bay; nlca40-teifi-valley; nlca42-pembroke-and-carmarthan-foothills; nlca44-taf-and-cleddau-vales; nlca46-preseli-hills; nlca48-milford-haven

These variations in content and formatting pose challenges from a text processing perspective. Parsing text from PDFs is a common headache in digital humanities research, but our task was complicated thanks to the diverse nature of the documents in the corpus, and the presentation of multimedia data in individual items. In terms of extracting the text from these PDF publications to build a corpus, we needed a tool that would be able to extract the prose from the LCAs as faithfully as possible, whilst also preserving the documents' material integrity.

Many PDF parsers are freely and sometimes (in terms of code) openly available. However, not all achieve the same performance or have the same features. We experimented with two main types of parser: one which could extract text on a page-by-page basis, and others which produce a structured output of the entire document (including heading labels and their content). Given that we were interested in the relationships between words, another desirable feature was dehyphenation of words broken over two lines. However, we only assessed readers freely available, and found that this feature is often offered in paid products only or versions that are not readily accessible. There were a total of 14 criteria against which we measured the PDF extractors[7], which can be classified into four main categories: global features (to features that are invariable from document to document), linear faithfulness, lossiness and structural oddities (Table 9.1).

We produced a sub-corpus especially for the task of assessing PDF extractors. Rather than choosing whole documents for this purpose, we sampled a number of pages from various LCAs. Only choosing pages in this manner meant that the annotation task could be done in a matter of days rather than weeks. In total, we had 19 PDFs in our sub-corpus, manually annotated for the 14 criteria.

Based on our criteria, XPDF emerged as the best choice of PDF text extractor for this project. It had the advantage of being able to render PDFs one page at a time natively, though it did not produce a structured output. While this parser was weakest in the rendering of tabular data and listed data, it was the most reliable in the areas most important for us given our intended analytical approach: it produced all content most reliably – it did not lose large sections of text, and reproduced highlighted text faithfully and, mostly, in situ. These strengths meant that it offered the most faithful rendering of the prose elements of the LCAs. Once we had produced these machine readable versions of the PDF documents, we could experiment with interdisciplinary, multiscalar approaches to analysing them.

---

[7] tika: https://pypi.org/project/tika/; pdfminer: https://pypi.org/project/pdfminer/; pdf2tree: https://pypi.org/project/pdftotree/; pdfquery: https://pypi.org/project/pdfquery/; xpdf: https://www.xpdfreader.com/download.html; pdfbox: https://www.xpdfreader.com/download.html

Global features	Linear faithfulness	Lossiness	Structural oddities
Structured output in XML or JSON format	Highlighted text rendered in proper sequence	Highlighted text is rendered (not necessarily in situ but at least reproduced)	In situ rendering of text for three-column formatted documents
Page-by-page processing	Non-column or highlighted related text rendered faithfully (e.g., table of contents)	Image text (e.g., the legend in a map) is rendered	In situ rendering of text for two-column formatted documents
	Image or caption text rendered in linear order (according to text flow)	Figures (non-map textual data) are rendered	Tabular data is rendered logically
		Major sections of text are not lost	Automatic reorientation of perpendicular text (i.e., not rendered one letter per line)
		Tabular data is rendered	

Table 9.1: Criteria for assessing PDF Extractors.

9.1.3   Multiscalar approaches to LCAs

Chesnokova et al. (2019) have demonstrated that a multiscalar approach is necessary for making meaningful contributions to the work of LCAs, and our approach here is in conversation with that earlier study. We focus on multiscalar approaches to text analysis taken from natural language processing (i.e., topic modelling) and literary studies (close reading). The process we describe blurs boundaries between quantitative/distant and qualitative/close methodologies. To perform our multiscalar text analyses we began with automatically generated topic models, close-read these results and then re-ran the models based on human interpretation of what 'value' signified in this corpus. One of the challenges of our multiscalar analysis is to create an iterative loop between human-led close reading and machine learning-led text analysis. By integrating the two approaches, we sought to treat the patterns generated through topic modelling as a guide for further research questions, rather than assuming that this human feedback can come at a later stage. Topic modelling was particularly appropriate for our work since it is an unsupervised approach well suited to an explorative quantitative analysis, and in our relatively small corpus it was straightforward to link emerging tokens of interest back to individual documents for a qualitative close reading.

Our approach uses elements from both ends of this spectrum: while our starting point is an unsupervised machine learning process, the interpretation of the results and subsequent re-runnings of the topic models apply more specific criteria in light of our findings uncovered through close readings of the documents. Not only did this method allow for more complex and flexible interpretations of the data, but it also echoed the interplay between the personal and the governmental – a social form of closeness versus distance in a given landscape – that shape understandings of 'value' in the LCAs. In this way, as we explore in the discussion below, we were able to develop multiscalar evaluations of the LCAs' implications for landscape management and cultural heritage. First, though, it is important to understand the two main processes – topic modelling and close reading – that constituted this particular form of multiscalar text analysis.

9.1.3.1   Topic modelling

In literary analysis terms, topic modelling is a form of distant reading; that is, it enables the 'tracing [of] a formal element through a vast body of works' and then attempts to 'build an explanatory model of the emergence, demise, or transformation of certain aspects' of the text (Khadem, 2012, p. 410). More specifically, topic modelling aims to discover the hidden semantic patterns across a document collection (Blei and Lafferty, 2009; Blei, 2012; Boyd-Graber, Mimno, and Newman, 2014) (see also Chapter 3). These patterns take the form of a probability distribution over the words in a training corpus, which is used to build the machine learning models.

There are many variants of the topic modelling algorithm and the one we employ is latent Dirichlet allocation (see Chapter 3 for more details). Once topics are generated for a particular document collection, we require a method of automatically evaluating these topics. There are a number of ways of ascertaining the quality of the learned topics, which involve *intrinsic* and *extrinsic evaluation* (Newman et al., 2010). Extrinsic evaluation integrates the output of the topic models as part of another task, with the assumption that the better the quality of the topics chosen, the better the external task will perform. Intrinsic evaluation assesses the topics in and of themselves. Intrinsic evaluation can involve *direct* and *indirect methods* (Lau, Newman, and Baldwin, 2014). One such indirect method involves the ability for humans or algorithms to detect intrusion words, or words that do not belong to a topic. In our work, we employ one of the direct methods in Lau et al. (2014), which calculates the observed coherence between the terms in a topic. The observed coherence measure is a means to determine how frequently words identified as belonging to the same topic occur together in a reference corpus.

Topic models are able to be coaxed in a particular direction during the learning process into favouring certain terms, called seed terms. In this way, they share with close reading a tendency to amplify certain terms or themes to aid the development of an argument.

9.1.3.2   Close reading

Close reading is a critical methodology that has been central to literary scholarship – the 'sine qua non of literary study,' according to Jonathan Culler (2010, p. 20) – for the last hundred years. It is closely linked with practical criticism and formalism, developed at Cambridge in the 1920s and 1930s by scholars including I. A. Richards, F. R. Leavis and William Empson. The aim, for these critics, was to get readers to attend to 'the words on the page' (Richards, 1929) and nothing else: a form of pure reading divorced from historical or contemporary contexts.

Modern literary studies inherit this focus on 'the words on the page', but is more generous about the extra material that can be brought to bear on the text. Close reading no longer operates in a scholarly silo; there are serious political, social and cultural implications from close reading, too. Henry Louis Gates Jr., for instance, viewed the close reading of black literature in the 1980s as being crucial for the sensitive recovery and acknowledgement of the repressions and cruelties that were figuratively or explicitly represented in a text (Gates Jr, 1990, p. 20). Reading the text on its own terms necessarily required the close reader to emphasise the unique conditions that each work – and each reading – represented.

For a time, a dichotomy seemed to be emerging between computational 'distant reading' (Moretti, 2013) and the close attention to detail with which literary scholars are more familiar. But both of these approaches, taken in isolation, have significant limitations for our understanding of textual sources; as Matthew Jockers wrote in 2013, 'The[se] two scales of analysis ... should and need to coexist' (Jockers, 2013, p. 9). Nowadays, most digital humanists and computational text scholars recognise that, in the words of Adam Hammond, Julian Brooke and Graeme Hirst, 'computational analysis can only thrive in an ecosystem of close reading' (Hammond, Brooke, and Hirst, 2016). Multiscalar analysis of the kind we describe here offers a 'flexibility' in reading and analysis that is not achievable in any single method (Taylor, Gregory, and Donaldson, 2018). This methodology, we would argue, is the new phase of the 'broad intellectual shift' that has transformed humanities and social science disciplines: multiscalar analysis recognises that previously discrete disciplines offer, in fact, interdependent ways of thinking about complex sources (Underwood, 2016).

From a literary studies perspective, any machine-led interventions need to be 'every bit as perspectival, multifaceted, and blurry' (Underwood, 2016) as a close reading can be. One of our aims here has been to respect the 'blurry' and subjective aspects of landscape valuation, alongside those facets of the LCAs that could be quantified through topic models [and sentiment analysis]. The result, as we demonstrate in the next section, is a multiscalar, multifaceted analysis that reflects the complexity of the LCAs' textualities in its methodological approaches.

9.2   Results and Discussion

9.2.1   Assessing "Value" by integrating knowledge and technology

Our first task was to determine what themes could be identified across the corpus of LCAs that could offer an insight into what aspects of the landscape might hold 'value' in this context. We first allowed a topic model to extract themes and issues without any human guidance. That is, topics were discovered and extracted rather than being selected a priori by us. We began with three runs of topic modelling, each producing 100 topics. A topic is represented by a collection of salient terms that co-occur together over and over again throughout the document collection. Examples of these extract topics are shown below:

- Topic 1: assessment character value identify coastal natural study environment information outstanding,
- Topic 2: site subdivision value heritage include natural feature character coastal building,
- Topic 3: applicant management issue project articulate resource land include section policy,
- Topic 4: development matter soil character use land change site large activity,
- Topic 5: change development activity legislation policy character natural manage provide farm.

From these initial discovered topics, we identified a series of seed terms in keeping with the themes and issues foregrounded by the texts themselves. We manually selected the topics which included the word 'value', and used the other terms in those topics to identify potential seed terms. For instance, in the examples above Topic 1 contains the word 'value', and thus the other terms in that line ('assessment', 'character', 'identify', 'coastal', 'natural', 'study', 'environment', 'information', 'outstanding') are all potential seed terms. By manually comparing the terms associated with 'value' in each of these topics, and collating the most common words (e.g., 'character') or themes (e.g., the environment), we identified a small sub-set of seed terms that collocated with the term 'value'.

In this way, we were able to identify a series of terms that were closely associated with any mention of 'value' within the landscape character assessment.

9.2.2   The value of character

Notwithstanding the LCAs' professed intent, our analysis made clear that there are certain character traits which are considered more valuable than others. Indeed, the word 'character' was one of the terms correlating most strongly with 'value'; they co-occurred together 24 times (almost double the average). This result, and a close reading of extracts from individual documents, undercuts the LCAs' claim that a character assessment 'seeks to capture baseline information about the character of the Park's landscapes in a value-free way' (Lakes), applying 'value' instead to the subsequent judgements as to the sensitivities of a landscape (North Norfolk). Yet, if 'character' and 'value' are so closely linked linguistically, can the documents really be said to be assessing character without value?

The North Norfolk LCA subtly highlights this tension:

> In considering landscape in land use planning and management, there has been a change in emphasis from landscape evaluation or designation, i.e. what makes one area 'better' than another, through to describing the 'character' of a landscape, i.e. what makes one area 'different' or 'distinct' from another.

The value judgement has not, in this new system, been eradicated; rather the capital by which value is assessed has altered. In these documents, what is valuable – what is singled out the primary defining feature of a character area – is *uniqueness*, or more precisely *distinctiveness* (the combination and interrelationship of a large number of features and components of a piece of territory which make it different to any other area). However, as we will see, the LCAs frequently borrow directly from each other, meaning that uniqueness is expressed in a rather homogenous way.

This repetition made assessing regional understandings of value difficult, since it was clear that the documents were responding carefully to centralised dictats. Of the four UK LCAs in which the connection between 'character' and 'value' was the most pronounced, three (Culm, Durham and Thames) state their aims in identical terms:

> Biodiversity and geodiversity are crucial in supporting the full range of ecosystem services provided by this landscape. Wildlife and geologically rich landscapes are also of cultural value and are included in this section of the analysis. This analysis shows the projected impact of Statements of Environmental Opportunity on the value of nominated ecosystem services within this landscape.

The implied goals, here, are twofold: to demonstrate the indivisible links between biological, geological and cultural richness; and to assess how new proposals will affect a given landscape. This, in a nutshell, is the means by which an LCA can measure its success: the extent to which the singularity of a place is defended against new policies that prioritise economic initiatives ('opportunity', in this sense, is a capitalist term) over 'ecosystem services'. To protect its region, an LCA must express its individuality in terms that are understandable to a central decision-maker. Uniqueness of landscape is, therefore, often obscured behind uniformity of expression.

LCAs implicitly agree on the main contributors to an area's character: which often translates into how rare its ecological elements are and how deep its history runs. For instance, the Devonshire and Durham grasslands are, geologically speaking, dissimilar – but they share a deep importance in the cultural character of their locations. The LCAs for both Durham and the Culm grasslands emphasise, particularly in relation to their ecological dimension, that these habitats are unique. The Culm LCA states that '[t]his habitat is unlike any other in England', and, indeed, is 'one of the last strongholds of rush pasture or Culm grassland in Britain'. The militaristic language – 'stronghold' – positions this landscape as the plucky survivor of a campaign against Britishness; its survival, then, is of interest not just to the landscape's character, but to national identity. The Durham LCA goes even further: it claims that this NCA hosts a globally 'unique' community of vegetation and invertebrates. To alter it is to destroy something that is not merely locally significant, but internationally valuable. These geological characteristics are only valuable, though, when combined with evidence of historical human action; the LCAs are united in believing that their distinctiveness arises from longstanding evidence of different relationships between people and place. Each articulates a sense that the landscape's character is an inviolable right that evidences British democracy from the ground up.

Intangible heritage, though, is not enough; an area's history must be 'visible', and in being put on display rendered 'timeless' (Culm LCA): as much a part of a landscape's present as its past. This visible history might take several forms: settlements (such as Clovelly, which the Culm LCA asserts is 'the first coastal settlement of which there is a firm record'), evidence of old industry (such as the 'allotments and pony paddocks, reclaimed colliery sites, disused and existing railways, and industrial archaeology' that pepper Durham), and remnants of agricultural practices, including patterns of field enclosures, hedgerows, sunken lanes and old farmsteads. The challenge, in locations like Culm and Durham, is to both 'protect and enhance' ecologically rich places for 'tranquillity and inspiration', and to 'illustrate' an area's human past (Durham LCA).

North Norfolk, on the other hand, faces a different problem: its historical cultural value has already been eroded, and the LCA demonstrates how severe the impact has been for the value of the landscape's character. This LCA utilises a grading system for 'Strength of Character', which is assessed as being 'poor', 'fair', 'moderate' or 'good' – and much of the landscape is found wanting. Evidence of modern lifestyles abounds in this area: the growth of small towns and villages by new housing developments, expanding suburbs, subdivided gardens, property extensions, barn conversions, agricultural buildings at a remove from farms, the renegotiation of field boundaries and reallocation of pasture, wind turbines, telecom masts, new roads (with attendant suburban features such as 'kerbing / signage / widening'), and the expansion of suburban traits into the open countryside ('surfaced drives / domestic style gates and fences / garden style planting and parking areas / overly large windows and external lighting, etc.') all have a detrimental impact on the landscape's character in ways that, as the LCA concludes, are 'individually modest but cumulatively significant'. The result of these 'erosion[s]' to the area's character is a landscape that feels 'somewhat degraded': its rich human and natural histories have been extensively buried beneath modern developments. Its value, consequently, has been dramatically reduced.

The core problem identified by the North Norfolk LCA is that its 'natural' character – that bestowed by features such as meadows, woodlands and floodplains – has been subdued by human interferences which risk homogenising the landscape. As the Durham LCA explains, some LCA consider that an area's uniqueness is determined by its 'natural' features; it is these that provide 'such [a] strong sense of place' (Durham LCA). This is a wider issue with LCAs: whilst accepting that landscape is dynamic and necessarily subject to change, and whilst treasuring the effects of change across time as long as it is not too recent, LCAs often fail to see any merit in 'modern' landscape change, which is generally seen more as 'erosion' rather than the creation of new forms of landscape character.

9.2.3   *Natural value and valuing nature*

This concern for non-human features is reflected in the list of topics that correlate with 'value'; the closest link is with the 'natural', which co-occurs with 'value' 26 times (more than double the average number of co-occurrences, and more frequently even than 'character'). This partly reflects the disciplinary backgrounds of some LCAs' authors, and also correlates with political factors: each LCA is aligned with Natural England, the government body whose stated aims are 'to secure a healthy natural environment for people to enjoy, where wildlife is protected and England's traditional landscapes are safeguarded for future generations' (LCAs - Culm, Durham). Even here, though, what 'natural' means is not neutral: certain kinds of 'natural' features are considered more valuable than others.

In the British LCAs, 'traditional' features associated with historic cultural and political identities are prioritised. Vegetation, particularly woodland, forests, trees, and hedgerows, and coastlines, including estuaries, seem, in our analysis, to be the most highly regarded 'natural' forms, as shown below in the seeds derived from the initial topic modelling runs.

> vegetation, woodland, tree, hedgerow, planting, forest
> natural, environment, character, assessment
> coastal, water, estuary

Close reading reveals that this phenomenon is partly attributable to the LCAs' detailed descriptions of local landforms, a focus on topographic form that derives from the geographical and landscape architect background of many LCA practitioners. Partly, these LCAs hark back to nostalgic notions of the past that link landscape aesthetics with the country's naval and imperialist histories, from the woodland heroics of Robin Hood to feats of naval conquest that were a source of national pride well into the 20th century. But there is another reason, too: on a densely populated island, areas that retain the peace and tranquillity of yesteryear are vanishingly few.

The 2007 Intrusion Map (CPRE)[8] demonstrated the extent to which urban development was intruding both visually and audibly on rural landscapes: that is, on those areas that had higher proportions of 'natural' features, and so are perceived as 'natural' landscapes. In the Culm LCA, that means that the 'simple, austere character of the landscape and seascape' should be preserved to protect the region's 'wealth': a non-economic form of riches that, the LCAs worry, are becoming rarified in the UK. As this LCA notes, the Intrusion Map suggests that the last strongholds of tranquillity are in the UK's woodlands. To this, the Durham LCA adds nature reserves and the coast.

The LCAs also find pressing practical reasons to prioritise the preservation of these tranquil areas. The Thames LCA explains:

> Protect and manage the area's historic parklands, wood pastures, ancient woodland, commons, orchards and distinctive ancient pollards, and restore and increase woodland for carbon sequestration, noise and pollution reduction, woodfuel and protection from soil erosion, while also enhancing biodiversity, sense of place and history.

This objective links that 'sense of place' and a landscape's heritage — tangible and otherwise — to concerns for the future. Without these 'historic', even 'ancient', woodlands the landscape's future looks bleak, because it is these characteristics that protect it from pollution, erosion and habitat loss. And these features have a significant role to play in safeguarding that other key element of the UK's national landscape character: its coastlines. The Lake District LCA summarises that the country's 'coastal margins are vulnerable to a range of climate change effects', including changing weather patterns, invasive species and the

---

8 https://www.cpre.org.uk/resources/intrusion-map-england-2007/

alteration (even destruction) of habitats. The result is not simply, as the LCA has it, effects on 'the character of the landscape'; the implication is that it risks the foundations of national character, too. To value a local landscape, then, stands in for valuing the whole country – but, there, the issues of how to express diversity in a homogenised language are even more vexed and pressing.

9.3   Conclusions and Further Work

Defining value in LCAs is a complex and nuanced task that requires analytic techniques sensitive to both objective or quantifiable features, and to affective or emotional markers in a given landscape. Combining computer-driven topic modelling with human-led close reading – and, crucially, forming an iterative loop between the two processes – has allowed us to get closer to an understanding of what we associate with 'value' in these documents. The results are perhaps not as straightforward as LCA authors might wish: the documents uphold a certain bias towards non-human elements of the landscape. Most of the time, this bias is important for maintaining an area's ecological health – but it does also risk sidelining human interests and suspending a landscape at a particular historic moment that may not match contemporary human concerns.

What those concerns are also change over time, and across geography. This study has raised further questions; for instance, what demographic assumptions are the LCAs making about the prime users of an area? What would happen to our understandings of 'value' if we were to investigate the relationship between 'value' and references to, and input from, indigenous populations? Can we delve deeper into the emotional value ascribed to a landscape, for instance, by applying sentiment analysis to these documents? Or could we gain a more nuanced understanding of a landscape's individuality by comparing these kinds of political documents with creative writing from a given area? Much remains to be done to unpack the assumptions and biases inherent to these documents, and to understand how these concealed facets – evident only through careful and multivalent analysis of the language used – affect landscape perception and management. Recognising LCAs as a type of environmental *narrative*, rather than straightforward reportage, uncovers a new area of research that will be crucial to navigating future steps in managing these landscapes for the future.

References

Blei, David M (2012). "Probabilistic topic models". In: *Communications of the ACM* 55.4, pp. 77–84. DOI: 10.1145/2133806.2133826.

Blei, David M and John D Lafferty (2009). "Topic models". In: *Text mining: Classification, clustering, and applications*. Ed. by A. N. Srivastava and M. Sahami. Boca Raton: CRC Press, pp. 71–94. DOI: 10.1201/9781420059458. ch4.

Bloomer Tweedale Architects and Town Planners (1992). *Mapping the Distribution of Special Landscape and Wildlife Areas Identified in Development Plans in Wales.*

Boyd-Graber, Jordan, David Mimno, and David Newman (2014). "Care and feeding of topic models: Problems, diagnostics, and improvements". In: *Handbook of mixed membership models and their applications.* Vol. 225255. Boca Raton: CRC Press.

Chesnokova, Olga, Joanna E. Taylor, Ian N. Gregory, and Ross S. Purves (2019). "Hearing the silence: Finding the middle ground in the spatial humanities? Extracting and comparing perceived silence and tranquillity in the English Lake District". In: *International Journal of Geographical Information Science* 33.12, pp. 2430–2454. ISSN: 1365-8816. DOI: 10 . 1080 / 13658816 . 2018 . 1552789.

CSA Environmental (2017). *Why is Landscape Value the topic of the day?* https://www.csaenvironmental.co.uk/2017/09/01/landscape-value-topic-day/.

Culler, Jonathan (2010). "Introduction: Critical paradigms". In: *PMLA* 125.4, pp. 905–915. DOI: 10.1632/pmla.2010.125.4.905.

Fairclough, Graham, Ingrid Sarlöv Herlin, and Carys Swanwick, eds. (2018). *Routledge handbook of landscape character assessment.* London and New York: Routledge. DOI: 10.4324/9781315753423.

Gates Jr, Henry Louis (1990). *Black literature and literary theory.* New York: Routledge.

Hammond, Adam, Julian Brooke, and Graeme Hirst (2016). "Modeling modernist dialogism: Close reading with big data". In: *Reading modernism with machines.* Berlin/Heidelberg: Springer, pp. 49–77. DOI: 10.1057/978-1-137-59569-0_3.

Institute, The Macaulay Land Use Research (1996). *Landscape Designations.* https://macaulay.webarchive.hutton.ac.uk/ccw/task-two/designations.html.

Jockers, Matthew L (2013). *Macroanalysis: Digital methods and literary history.* Champaign: University of Illinois Press. DOI: 10 . 5406 / illinois / 9780252037528.001.0001.

Khadem, Amir (2012). "Annexing the unread: A close reading of "distant reading"". In: *Neohelicon* 39.2, pp. 409–421. DOI: 10.1007/s11059-012-0152-y.

Landscape Institute (2013). *Green Infrastructure: An Integrated Approach to Land Use.* https://landscapewpstorage01.blob.core.windows.net/www-landscapeinstitute-org/2016/03/GreenInfrastructureLIPositionStatement2013.pdf.

Lau, Jey Han, David Newman, and Timothy Baldwin (2014). "Machine reading tea leaves: Automatically evaluating topic coherence and topic model quality". In: *Proceedings of the 14th Conference of the European Chapter of the*

Association for Computational Linguistics, pp. 530–539. DOI: 10.3115/v1/ E14-1056.

Ministry of Housing, Communities and Local Government (2019). *National Planning Policy Framework.* https://assets.publishing.service.gov.uk/ government/uploads/system/uploads/attachment_data/file/810197/NPPF _Feb_2019_revised.pdf.

Moretti, Franco (2013). *Distant reading.* London: Verso Books.

Newman, David, Jey Han Lau, Karl Grieser, and Timothy Baldwin (2010). "Automatic evaluation of topic coherence". In: *Human language technologies: The 2010 annual conference of the North American chapter of the association for computational linguistics*, pp. 100–108. DOI: 10.5555/1857999. 1858011.

Office, Scottish (1996). *Natural Heritage designations review: discussion paper.*

Richards, Ivor Armstrong (1929). *Practical criticism: A study of literary judgment.* Whitelackington: Kegan Paul.

Taylor, Joanna E, Ian N Gregory, and Christopher Donaldson (2018). "Combining close and distant reading: A multiscalar analysis of the English Lake District's historical soundscape". In: *International Journal of Humanities and Arts Computing* 12.2, pp. 163–182. DOI: 10.3366/ijhac.2018.0220.

Underwood, Ted (2016). "Distant reading and recent intellectual history". In: *Debates in the Digital Humanities*, pp. 530–533. DOI: 10.5749/j.ctt1cn6thb. 47.

CHAPTER 10

Interpreting Natural Spatial Language in a Fictional Text: Analysing Natural and Urban Landscapes in Mary Shelley's *Frankenstein*

Tobias Zuerrer

University of Zurich, Switzerland

This chapter explores how landscape is described in Mary Shelley's *Frankenstein*, a novel which was first published in 1818. The book emerged at a time when English society was transforming from a predominantly rural to an industrial one. While this transition led to technological progress, people simultaneously turned back to nature as a source of inspiration and refuge from the city (Squire, 1988). The natural world was thus often seen as a place of tranquility, tucked away from the daily pressures of industrial society.

Written texts, in the form of blogs, tweets, Facebook posts or other documents can also serve as tool for geographers interested in how individuals (of the past) perceive(d) and respond(ed) to the landscapes around them. As Chesnokova and colleagues note, 'written accounts can provide us with one – admittedly incomplete, yet nevertheless significant – way of understanding what people describe when communicating about both contemporary and

How to cite this book chapter:
Zuerrer, Tobias (2022). "Interpreting Natural Spatial Language in a Fictional Text: Analysing Natural and Urban Landscapes in Mary Shelley's *Frankenstein*." In: *Unlocking Environmental Narratives: Towards Understanding Human Environment Interactions through Computational Text Analysis*. Ed. by Ross S. Purves, Olga Koblet, and Benjamin Adams. London: Ubiquity Press, pp. 197–210. DOI: https://doi.org/10.5334/bcs.j. License: CC-BY 4.0

historical landscapes' (Chesnokova et al., 2019, p. 2432). Textual analysis is an important ancillary to understanding how individuals' perception of and reaction to landscapes varies across time and space. Contemporary linguistic research has increasingly seen fictional language as a not 'artificial, deficient or contrived [but] as a rich source of data, albeit one that needs to be investigated on its own terms' (Locher and Jucker, 2017, p. 5). For linguists interested in how individuals of the past viewed the world around them, fictional texts are a gateways to these perceptions. Historical linguists argue that even though fictional language deviates from everyday, spontaneous interactions, 'such material […] is the only possible source for the spoken language of the past' (Jucker and Taavitsainen, 2010, p 8).

In terms of setting, the story *Frankenstein* takes place in central Europe as well as the Arctic region. The story begins with the frame narrative of Captain Robert Walton, who finds himself on journey to the North Pole. On this voyage, the captain encounters Victor Frankenstein who subsequently begins to tell his life story. A few years back, in an attempt to create life from individual body parts, Frankenstein had succeeded in bringing into existence a live 'creature'. As the reader learns, this creature had gone on to destroy Frankenstein's life. In an act of vengeance, the latter chases his creation from Geneva, Switzerland, all the way to the Arctic circle. On their journey, the protagonists travel through different landscapes, from cities to villages, through rugged mountain ranges and across lakes. The natural as well as the urban world plays an important role in the story and is referred to numerous times.

As a first step, this project investigates the ways in which geographical features/toponyms of the 'natural' and the 'urban' landscape are conceptualised in *Frankenstein* and whether, and to what extent, one can find a dichotomy between the two. In a second step, based on the findings, it will be discussed, whether a fictional work like *Frankenstein* is a suitable source to investigate the way in which a landscape was perceived by people in the past.

10.1   Documenting the Perception of a Landscape Using Text

Several studies have used text as a source to generate spatial data of how landscapes (and prominent features within these landscapes) are described and perceived by individuals.

In a study characterising landscape variation, Derungs and Purves (2016) show how local folksonomies can give an insight into how landscapes are described over time and space. Through these folksonomies, for instance, policy makers can get a bottom-up understanding of how landscapes are conceptualised. In their investigation, they work with the Text+Berg and HIKR corpora – two online bodies of text that cover mountaineering activities in Switzerland. Their approach is a novel one in that it takes 'full text corpora as a starting point to generate rich, spatially referenced, landscape descriptions'

(Derungs and Purves, 2016, p. 69). This allows the researchers to identify different descriptions of landscapes across Switzerland at small scale.

In a further study looking into spatial expressiveness of landscape features, Derungs and Samardžić (2018) focus on one prominent feature of Switzerland's landscape. They focus on unique mountain names that are mentioned in a large collection of text on Swiss alpine history. They show that '[s]mall spatial extents, found all over Switzerland, can show considerably strong correlations between text frequency and spatial prominence' (Derungs and Samardžić, 2018, p. 856). While they do not focus on qualitative properties of certain landscape features, they illustrate that the prominence of a certain mountain is reflected in the number of times it is mentioned in a text.

As part of their project on identifying how individuals perceive silence in the landscape around them, Chesnokova et al. (2019) investigate historical and contemporary texts describing the Lake District in England. They do not study fictional texts contained in the historical Corpus of Lake District Writing but work with material that dates from the same decade as the present object of study. In a first step, the authors extract text segments that describe how tranquillity is experienced in the Lake District. In a subsequent step, the study explores how different terms related to silence are interpreted differently over the decades. They find that 'what is meant by quietness has undergone a significant shift in the intervening years [...]' (Chesnokova et al., 2019, p. 2443). However, the authors do not only investigate how silence is described but also where it is spoken about. They find that a lot of descriptions refer to the area of Grasmere, an area that was first popularised by the romantic poet Wordsworth.

In a study looking into how different landscapes in Switzerland are described, Wartmann et al. (2018) find that descriptions of a certain place or region vary depending on which source text is used. They find for instance that Flickr tags contain a lot of toponyms whereas entries in hiking blogs offer more information on the sense of a place. This results in 'descriptions from the same data source [being] more similar than between data sources, also for different landscape types' (Wartmann, Acheson, and Purves, 2018, p. 1585). Therefore, different places in Switzerland are often described in a similar fashion by a certain text source. In contrast, the same area in Switzerland is often described differently by texts from different sources. As the scholars emphasise, this does not need to be regarded as a disadvantage. Rather, the varying emphases on what is described provide the researcher with multiple perspectives on how a landscape is viewed.

For the present investigation, which is interested in the landscape described in Frankenstein, the studies outlined above will be taken into consideration. It will be interesting to see whether one is able to come up with a certain characterisation of the landscape, based on the words that appear in conjunction with a seed term. Like Derungs and Samardžić (2018) find, a prominent mountain of the Alps is mentioned quite frequently in the text. Mont Blanc is mentioned eight times and it will be interesting to see, the way it is referred to. While it will not be possible to do a temporal comparison between *Frankenstein* and a more recent novel (similar to the approach taken by Chesnokova et al. (2019)), this study will investigate how geographic features and toponyms in a certain place are referred to. As Wartmann et al. (2018) emphasise, it needs to be considered that contributors to and authors of different texts resort to different strategies when describing a landscape. It is to be expected that a novel takes a different approach to describing a landscape than a hiking blog or a Facebook post. Nevertheless, a novel such as *Frankenstein* might offer an additional insight to how a certain landscape is perceived and talked about in contemporary culture.

10.2   Methodology

The novel *Frankenstein* is freely accessible online and can be downloaded as a text file. Using AntConc, 'a freeware corpus analysis toolkit for concordancing and text analysis' (http://www.laurenceanthony.net-/software/antconc/), a word list was generated. Using this list, a set of geographical features and toponyms were extracted. This selection was arbitrary and is based on the author's preconceptions. In a next step, similar to the approach taken by Chesnokova et al. (2019), the Historical Thesaurus of English was consulted to search for historical synonyms of the seed words selected. However, this step proved unnecessary as no relevant synonyms could be found. Subsequently, the seed terms were placed into two categories; toponyms / geographical features belonging either to the 'natural' or the 'urban' landscape. The individual seed terms as well as the number of hits retrieved are shown in Table 10.1.

There is a total of 138 occurrences where the seed terms related to natural landscapes are mentioned. For search terms attributed to the urban landscape, a total of 100 hits were retrieved. The term 'mountain' is by far the most prominent geographical natural landscape feature in the text. This can be explained by the setting of the story as well as the journey that the main characters take. 'Geneva'

Natural landscape		Urban landscape	
Seed term	**Hits**	**Seed term**	**Hits**
mountain*	58	Geneva*	36
lake*	32	town*	30
river*	17	village*	15
glacier*	8	cit*	14
Mont Blanc*	8	Chamounix	5
Total	123	Total	100

Table 10.1: Seed terms (note use of wild cards) retrieved in Mary Shelley's *Frankenstein*.

is the most frequently mentioned toponym. While only a small part of the story takes place in the city itself, it is the birthplace of the narrator and plays an important role in his development as a scientist.

In a next step, AntConc's concordance tool was used to investigate the context in which the seed terms appear. Each occurrence was analysed in isolation. Firstly, it was decided whether the respective term was being qualitatively described or not. If the former was the case, the term was characterised as being either positively, neutrally or negatively connotated. Examples for each case are shown below.

1. **No categorisation possible:**
   [...] for the birth of that passion which afterwards ruled my destiny I find it arise, like a **mountain** river, from ignoble and almost forgotten sources [...](Shelley, 2012, p. 68).
2. **Positive connotation:**
   [...] the sublime shapes of the **mountains**, the changes of the seasons, tempest and calm, the silence of winter, and the life and turbulence of our Alpine summers [...] (Shelley, 2012, p. 320).
3. **Neutral connotation:**
   [...] a peaked **mountain** to the east of the lake (Shelley, 2012, p. 99).
4. **Negative connotation:**
   Immense and rugged mountains of ice often barred up my passage [...] (Shelley, 2012, p. 225).

In example (1) the seed term 'mountain' cannot be categorised for two reasons. Firstly, the term forms a part of the compound noun 'mountain river'. The author refers to a flowing body of water rather than the mountain itself. Secondly, 'mountain river' is employed as a simile. What is being described is the protagonist's passion and not the physical object itself. In example (2) the narrator describes the lofty shape of the mountains, which are portrayed in a positive light. In contrast, the term 'mountain' was seen to be neutrally described in example (3). Here the adjective 'peaked' merely describes the form of the mountain and, based on the context, there is no evaluative judgement. In the last example, the mountains are portrayed negatively. They are not shown to be beautiful but rather as an obstruction which hinder the protagonist from moving onward.

In some instances, the categorisation proved quite difficult. It lies within the nature of fictional works that a certain amount of interpretation is required by the reader. While reader A might interpret a text passage in one way, reader B might see that passage in completely different light. Take, for instance, the sentence below:

5. The desert mountains and dreary glaciers are my refuge (Shelley, 2012, p. 119).

According to the Oxford English Dictionary, the meaning of the word 'dreary' has been used in a similar fashion over the past two centuries and can be equated to 'uninteresting', 'repulsively dull' or 'horrid'. Nevertheless, in the instance above, the term 'glacier' was interpreted to be positively described. Following his expulsion from society, Frankenstein's monster views the glaciers as his refuge, and he goes on to state that the caves of ice are a dwelling to him. He sees them as a source of comfort. Nevertheless, other readers might read the sentence above differently.

10.3   Results

What can be seen from the examples above is that even with one seed term, the descriptions regarding that term can vary greatly. When analysing a seed term used in context, it is often not clear whether it is described positively or negatively.

Table 10.2 provides an overview of the way in which the seed terms related to natural landscapes have been categorised in the present study. In total, in around 70% of all mentions of natural landscape features, these features were either described as having positive neutral qualities. Only in 11.4% of all instances were these features described negatively.

From the seed terms related to nature, there was only one which was unanimously described in a positive way. Mont Blanc is alternatively described as 'supreme', 'magnificent' and 'beautiful'. In some cases, the narrator describes parts of the mountain such as its 'bright summit'. In other instances, the emphasis lies on the interaction between the surrounding environment and the mountain, that is, the way in which the lightning plays on the summit 'in the most beautiful figures'. The other seed terms are predominantly described in a positive or neutral fashion except for 'mountain' which forms a small exception. In around 17% of all instances, this natural landscape feature is described negatively. Mountains are described as 'inaccessible', 'immense' and 'rugged', barring the characters from climbing them or passing through them. In other cases, the

Seed term	Positive	Neutral	Negative	Other	Total
mountain*	25 (43.1%)	13 (22.4%)	10 (17.2%)	10 (17.2%)	58 (100%)
lake*	14 (43.7%)	10 (31.3%)	1 (3.1%)	7 (21.9%)	32 (100%)
river*	4 (23.5%)	6 (35.3%)	2 (11.8%)	5 (29.4%)	17 (100%)
glacier*	3 (37.5%)	3 (37.5%)	1 (12.5%)	1 (12.5%)	8 (100%)
Mont Blanc*	8 (100%)	0 (0%)	0 (0%)	0 (0%)	8 (100%)
Total	54 (43.9%)	32 (26.0%)	14 (11.4%)	23 (18.7%)	123 (100%)

Table 10.2: Seed terms related to natural landscapes.

mountains are shown to negatively affect the mood of the protagonists. In one instance, the narrator deems the perpendicularity of the mountain to create 'a scene terrifically desolate'. In another, the summits of the mountains 'hid in uniform clouds' create a 'negative mood'.

Table 10.3 shows the seed terms related to the urban landscapes. In contrast to what was expected at the start of this research project, the features of the urban landscape are described in a predominantly neutral and/or positive fashion. In a further third of all instances, a categorisation of the seed term was not possible. Interestingly, only two instances were noted where an urban landscape feature was described negatively. Possible reasons for this will be analysed in further detail in the discussion section.

The bar charts shown in Figures 10.1 and 10.2 graphically show the distribution of how the individual landscape features of both the natural and the urban world were described.

Seed term	Positive	Neutral	Negative	Other	Total
Geneva*	10 (27.8%)	0 (0.0%)	14 (38.9%)	12 (33.3*%)	36 (100%)
town*	8 (26.7%)	2 (6.7%)	14 (46.7%)	6 (20.0%)	30 (100%)
village*	4 (26.7%)	0 (0.0%)	6 (40.0%)	5 (33.3%)	15 (100%)
cit*	7 (50.0%)	0 (0.0%)	2 (14.3%)	5 (35.7%)	14 (100%)
Chamounix	2 (40.0%)	0 (0.0%)	1 (20.0%)	2 (40.0%)	5 (100%)
	31 (31.0%)	2 (2.0%)	37 (37.0%)	30 (30.0%)	100 (100%)

Table 10.3: Seed terms related to urban landscapes.

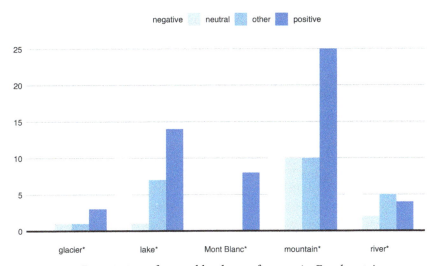

Figure 10.1: Description of natural landscape features in *Frankenstein*.

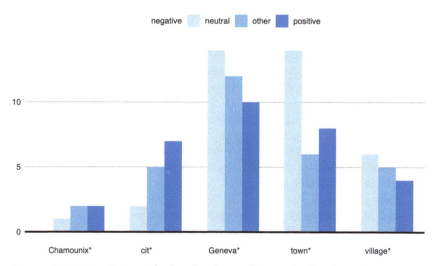

Figure 10.2: Description of urban landscape features in *Frankenstein*.

As the results show, based on the seed terms that have been chosen for analysis, there does not seem to be a clear dichotomy between the urban and the natural landscape apart from the fact that features of the natural landscape are often described more negatively whereas features of the urban landscape are described more neutrally. Again, it needs to be stressed here that this observation is based on a few arbitrarily selected seed terms. A more substantial investigation with more features might yield different results and distributions. Nevertheless, the next section will look more closely at individual instances to find possible explanations for the phenomena shown above.

10.4   Discussion

10.4.1   *Natural landscape features in* Frankenstein

Although it is very hard to generalise based on the sample that was taken, it appears that more prominent and inhospitable features in the landscape of *Frankenstein's* world are subjected to lengthier and more extreme descriptions, whether these be positive or negative. Consider the examples below.

6. The immense mountains and precipices that overhung me on every side
   […] and I ceased to fear or to bend before any being less almighty than
   that which had created and ruled the elements, here displayed in their
   most terrific guise (Shelley, 2012, p. 107).

7. I remembered the effect that the view of the tremendous and ever-moving glacier had produced upon my mind when I first saw it. It had then filled me with a sublime ecstasy that gave wings to the soul [...] (Shelley, 2012, p. 116).

8. [...] and at a distance, surmounting all, the beautiful Mont Blanc, and the assemblage of snowy mountains that in vain endeavour to emulate her [...] (Shelley, 2012, p. 196).

The examples above illustrate the fact that in some instances, mountains, glaciers and Mont Blanc are described in grand terms by the narrators. It goes without saying that there are numerous cases where these landscape features are merely mentioned or serve as reference points. However, in other cases there is a direct link between the features and the effect they have on the beholder. In example (6), given their immensity, the mountains are shown to be almost other-worldly, created by a being mightier than anyone else in the world. In extract (7), the narrator is left ecstatic at the sight of the immense ice before him. In the next example, Mont Blanc is singled out as the most beautiful of all mountains. However, the latter are also personified and infused with life.

In contrast to the findings of Derungs and Samardžić (2018) the prominent features listed above are not mentioned more frequently than other landscape features in the text. However, the author invests more text to not only describe the features in detail but to also explain the feelings that these features evoke in the characters and eventually the reader.

In the example below, different lakes are described by the narrator. The bodies of water are described as having a calm and soothing effect on the beholder. In example (9), the lake is shown to have charming quality. Its presence does not provoke excitement but rather a quiet admiration. Similarly, in examples (10) and (11) the still waters provide a feeling of calm. Words like 'ecstasy', 'sublime' or 'tremendous' are absent.

9. [...] but there is a charm in the banks of this divine river that I never before saw equaled (Shelley, 2012, p. 166).

10. [...] the sky and lake are blue and placid (Shelley, 2012, p. 98).

11. I contemplated the lake, the waters were placid; all around was calm (Shelley, 2012, p. 97).

Based on the seed terms chosen one can only speculate. However, it is possible that a lake is seen as a less prominent feature and does not need to be described in as much detail as the more inaccessible landscape features. Alternatively, it is to be expected that the sight of a calm lake does not produce the same excitement in the beholder. It is something that the viewer encounters on a daily basis.

10.4.2   Urban landscape features in Frankenstein

As mentioned in the discussion section, the urban seed terms analysed are sel-dom described in a negative manner. In a third of all instances, towns and cities are mentioned in passing, serving as a point of reference or as a stopover on the character's journey. This is shown in extract (12), where the reader does not learn anything about the qualitative properties of the city other than the fact that it offers protection to the woman in question. Rather, he/she learns about the location of the town in relation to a cottage that was previously mentioned. Similarly, in example (13) the narrator explicitly states that he does not know where he is passing through and as a result it is difficult to geographically posi-tion the town within the fictional world.

12. She arrived in safety at a town about twenty leagues from the cottage of De Lacey [...] (Shelley, 2012, p. 141).
13. I did not know the names of the town that I was to pass through [...] (Shelley, 2012, p. 152).

In the two extracts below, both Edinburgh and Paris are discussed. The two cities are alternatively described as having a regularity about them and as being both luxurious and beautiful. They fill the narrator with delight.

14. But the beauty and regularity of the new town of Edinburgh, its roman-tic castle and its environs, the most delightful in the world [...] (Shelley, 2012, p. 171).
15. A few months before my arrival they had lived in a large and luxurious city called Paris, surrounded by friends and possessed of every enjoy-ment which virtue [...] (Shelley, 2012, p. 137).

Although more seed terms related to the urban landscape need to be analysed, the cities described here are not described negatively. Rather they are shown to be a place of comfort and refuge. They do not evoke the same strong emotions as do some of the natural landscape features. In some instances, they merely act as signposts on the narrator's journey and the qualities are not directly described.

10.4.3   The natural and the urban – a landscape dichotomy?

Based on the investigations above, a mapping of the precise physical location of the landscape features is not always possible. Unlike Derungs and Purves (2016) who use the Text+Berg and HIKR corpus to allocate natural features to grid cells which are placed on a map, this study is unable to do that with the information from the novel. Often, there is no reference to the precise locations of the features in the landscape. However, as mentioned at the beginning, this

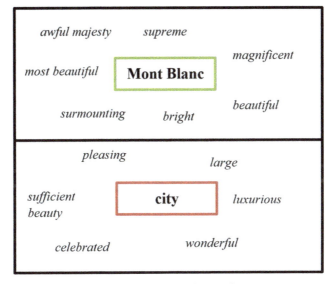

Figure 10.3: Spatial word clouds of two landscape features.

paper is not interested in the precise location of landscape features. Rather, its aim is to analyse the way in which certain places are described.

Similar to Derungs and Purves (2016) who create spatial word clouds that represent the most prominent landscape features in a certain grid cell in Switzerland, one could create a semantic word cloud for individual landscape features in Frankenstein. An example of such a cloud is presented in conjunction with 'Mont Blanc' and 'city'. Here, some of the words in the immediate vicinity (seven words to the right and seven words to the left) of the landscape feature were considered.

It is possible that further seed terms (in combination with words that appear in their vicinity) might show a dichotomy between how the natural and the urban landscape are described in the novel. However, if one looks at Figure 10.3, this dichotomy is not evident.

10.5   Further Studies

As Wartmann et al. (2018) note, the data source can have a significant impact on the way in which a certain landscape is described. Whereas hiking blogs might be more subjective, other sources are more factual. Although the world of *Frankenstein* is entirely fictional, there are different sources in the story from which the reader gets his/her information about the course of events. At the beginning and the end, the story is told from the perspective of Captain Walton. Subsequently, both Victor Frankenstein and the creature share their perspective on the course of events. The latter two find themselves in ever-changing situations and are shown to pass through states of distress and relative calm.

As a result, the three narrators have different relationships to the people and the landscapes around them. Walton is an avid explorer who wants to reach the North Pole at all costs. Frankenstein is a scientist who appreciates nature but also feels at home in the city. The creature is a social outcast who, as the story progresses, seeks refuge in the wild. When analysing a hiking blog, one can assume that individuals who contribute to the blog have a similar interest. In contrast, a novel like *Frankenstein* offers a blend of different views even though it is written by one and the same author. Whereas a glacier or a mountain might be seen as an obstacle by the explorer, the same landscape feature might be described as a refuge from the perspective of the creature and a sublime sight by Frankenstein.

In a novel it is thus difficult to generalise how a certain landscape (feature) is perceived. Although more novels would need to be analysed, it can be presumed that the author describes the surroundings so that they fit the storyline as well as the characters. Although a novel written in the Romantic era places a lot of emphasis on the natural world, this world is often the reflection of a certain character's mood. As a geographer, one cannot presume that this is the way the author thought of the landscape around him/her – rather he/she evokes a landscape that fits the narrative.

In a future study, it could be interesting to take into account the different narratological levels. In addition, it would be interesting to move beyond the fictional world and investigate how contemporary readers respond to the descriptions in the text and whether they share similar views to those expressed therein.

10.6   Conclusion

This brief study sought to investigate, whether there is a dichotomy between the natural and the urban landscape in Mary Shelley's *Frankenstein*. A small selection of landscape features has shown that these tend to be described more elaborately, the more prominent and/or inhospitable they are. However, there does not seem to be a clear difference in the description of natural and urban landscape features. A description of a lake was seen to be quite similar to that of a city. The body of water and the regularity of city were seen to evoke a sense of calm in the narrators. An investigation of more seed terms and/or novels might help paint a clearer picture.

References

Chesnokova, Olga, Joanna E. Taylor, Ian N. Gregory, and Ross S. Purves (2019). "Hearing the silence: Finding the middle ground in the spatial humanities? Extracting and comparing perceived silence and tranquillity in the English Lake District". In: *International Journal of Geographical Information Science* 33.12, pp. 2430–2454. ISSN: 1365-8816. DOI: 10 . 1080 / 13658816 . 2018 . 1552789.

Derungs, Curdin and Ross S Purves (2016). "Characterising landscape variation through spatial folksonomies". In: *Applied Geography* 75, pp. 60–70. DOI: 10. 1016/j.apgeog.2016.08.005.

Derungs, Curdin and Tanja Samardžić (2018). "Are prominent mountains frequently mentioned in text? Exploring the spatial expressiveness of text frequency". In: *International Journal of Geographical Information Science* 32.5, pp. 856–873. DOI: 10.1080/13658816.2017.1418362.

Jucker, Andreas H and Irma Taavitsainen (2010). *Historical pragmatics*. Vol. 8. Berlin: Walter de Gruyter. DOI: 10.1515/9783110214284.

Locher, Miriam A and Andreas H Jucker (2017). *Pragmatics of fiction*. Vol. 12. Berlin: Walter de Gruyter GmbH & Co KG. DOI: 10.1515/9783110431094.

Shelley, Mary (2012). *Frankenstein*. Peterborough: Broadview Press.

Squire, Shelach J (1988). "Wordsworth and Lake District tourism: Romantic reshaping of landscape". In: *Canadian Geographer/Le Géographe canadien* 32.3, pp. 237–247. DOI: 10.1111/j.1541-0064.1988.tb00876.x.

Wartmann, Flurina M, Elise Acheson, and Ross S Purves (2018). "Describing and comparing landscapes using tags, texts, and free lists: An interdisciplinary approach". In: *International Journal of Geographical Information Science* 32.8, pp. 1572–1592. DOI: 10.1080/13658816.2018.1445257.

Discovering Spatial Referencing Strategies in Environmental Narratives

Simon Scheider

Department of Human Geography and Spatial Planning, Utrecht University, the Netherlands

Ludovic Moncla

Univ Lyon, INSA Lyon, CNRS, UCBL, LIRIS, UMR5205, F-69622, France

Gabriel Viehhauser

Department of Digital Humanities, University of Stuttgart, Germany

In an environmental narrative, authors describe their perception and experience of a landscape in a form that enables the reader to follow their track in mind. The narrative resembles a consecutive set of snapshots of space viewed from a particular angle, either from an imagined ego on the track or from other locations specified relative to identifiable landmarks. This allows a reader to embed the authors' journey into the landscape, even if this landscape or journey are merely imagined (Tuan, 1991).

A key problem in analysing and comparing such narratives is the ability to *geo-reference* the places referred to in the text (Scheider and Purves, 2013)

How to cite this book chapter:
Scheider, Simon, Ludovic Moncla, and Gabriel Viehhauser (2022). "Discovering Spatial Referencing Strategies in Environmental Narratives." In: *Unlocking Environmental Narratives: Towards Understanding Human Environment Interactions through Computational Text Analysis*. Ed. by Ross S. Purves, Olga Koblet, and Benjamin Adams. London: Ubiquity Press, pp. 211–232. DOI: https://doi.org/10.5334/bcs.k. License: CC-BY 4.0

(cf. the chapter about spatio-temporal linking of narratives by Tim Baldwin in this volume). Geo-references not only allow texts to be put on a map but also to be segmented into situated scenes. Furthermore, places can be linked across documents, so that it becomes possible to track a given environment across different perspectives, diverse authors, or even different literary periods. For instance, suppose we would like to assess the environmental change of a landscape such as the *Moors in Scotland*, based on comparing narrative settings of historical novels with contemporary travel literature. To do this, we need to know when narrators talk about the same place.

One of the major methodical challenges is that environmental narratives seldom refer to the environment in terms of place names. Rather, spatial references are often *indirect*, that is, relative to perceived objects, and the difficulty is that automated *geocoding* tools and Natural Language Processing (NLP) strategies currently struggle with any references beyond place names (Purves et al., 2018; Chen, Vasardani, and Winter, 2018; Stock, 2014; Scheider and Purves, 2013). The diversity of strategies narrators have at their disposal to refer to a location have been subject of empirical study by cognitive linguists and anthropologists such as Levinson (Levinson, 2003) or Palmer (Palmer, 2002) across different language communities. A *frame of reference (FoR)* is a strategy for describing a given location relative to diverse sets of objects, including the perceiving ego or salient landmarks. Understanding this strategy is needed for a reader to comprehend the meaning of diverse locative expressions, ranging from egocentric ones such as 'the mountain in front of me' to allocentric ones such as 'the place where the river Ba flows into Loch Laidon', and from rather precise absolute references, such as 'ten miles north-east of Loch Laidon', to relatively vague descriptions without any directional hint, such as 'away from the waterway'. Frequently, authors of narratives also speak about a location only in a metaphorical way (Talmy, 1996), using *fictive motion* to move an imagined ego through a landscape. This is reflected, for example, in expressions such as 'the trail runs along the lake'. Also, temporal references can play an important role in spatial referencing (Tenbrink, 2011).

How can we discover FoRs in environmental narratives? The relevance of FoR for spatial referencing and geographic knowledge discovery has been known for a long time (Mark et al., 1999; Burenhult and Levinson, 2008). Qualitative models of spatial information were developed in the past precisely with an eye on such cognitive frames of reference (Clementini, Di Felice, and Hernández, 1997). From the viewpoint of geographic information retrieval (GIR) (Purves et al., 2018), the task of discovering FoRs in narratives has only been looked at sporadically in the past. It is apparent that this requires more than just building formal FoR models (Clementini, 2013), extracting parts-of-speech (PoS), spatial relation words without context (Stock and Yousaf, 2018), or the recognition of named entities (NER). What is needed includes, to the very least:

1. *Extracting those PoS* from a text that are needed for identifying the type of FoR
2. *Identifying the referencing strategy* (type of FoR + parameters) used by the speaker
3. *Georeferencing the parameters* used in the FoR
4. *Transforming the target location* into coordinate space, taking account of vagueness

While some research has recently been done to address the latter two challenges (Chen, Vasardani, and Winter, 2018; Scheider et al., 2018; Stock and Yousaf, 2018), the first two challenges about geoparsing are seldomly taken into focus (Moncla et al., 2014; Vasardani et al., 2012; Stock and Yousaf, 2018). In particular, it is still unclear which kinds of reference strategies need to be distinguished for environmental narratives, and to which degree they can be extracted from texts based on state-of-the-art geoparsing methods.

In this chapter, we illustrate how FoRs can be automatically discovered in Scottish narratives, and we test the quality of such discovery. We first explore a range of referencing strategies which occur in environmental narratives, without any pretence at completeness. We then assess how well both human annotators and geoparsers can be used to discover these strategies in three sample texts. Our goal is to support people interested in automated alignment and mapping of environmental narratives beyond place names. We will finally discuss to what extent the method is useful for this purpose.

11.1 Mountaineering in Scotland

As a literary basis for exploration, we selected two mountaineering texts which are narrative descriptions of a given landscape, namely *Rannoch moor* in Scotland. W. H. Murray's 1957 book *Undiscovered Scotland* talks about a hike through the moor in Chapter 17 'The Moor of Rannoch' (Murray, 2003). Fifty years later, in 2007, R. Macfarlane describes a similar trip through the moor in his book *The Wild Places* (Macfarlane, 2008, p. 73). While fictional texts do not necessarily aim at an explicit description of their settings, but rather evoke them implicitly (Viehhauser and Barth, 2017), *non-fictional travel literature* may be more likely to contain sophisticated references to actual landscape in a way that, we believe, reflects the diversity of spatial referencing in narrative texts. The texts serve us both as a source for discovering FoR diversity, as well as a source for evaluating the quality of annotation and geoparsing. For external validation of geoparsing, we used in addition a *fictional* text that describes a travel through Scotland, namely R. L. Stevenson's *Kidnapped*[1]. The intention is

---

[1] https://www.gutenberg.org/files/421/421-0.txt

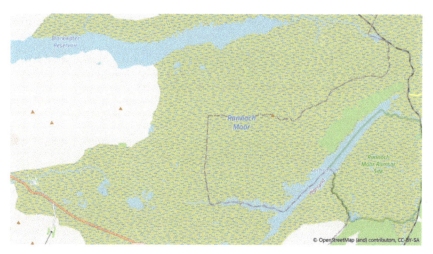

Figure 11.1: Rannoch Moor on Open Street Map (OSM). © Open Street Map contributors.

to identify the right referencing strategy used in these texts, in order to approximate the localisation of the many implicit places mentioned in the trip. While the latter are localisable only relative to toponyms such as 'Loch Laidon' and 'Loch Ba', the toponyms themselves can be easily georeferenced with standard geodata sources such as Open Street Map (Figure 11.1).

11.2   Referencing Strategies in Environmental Narratives

Within the sample mountaineering texts, we first explored the contained locative expressions (Herskovits, 1985), interpreting them in terms of known frames of reference or geometric strategies. In doing so, our intention was to capture the particular referencing strategy that might be used to technically reconstruct or approximate the referenced location in geographic coordinate space (c.f. Stock, 2014).

Types of FoR were proposed in Levinson (2003), Pustejovsky, Moszkowicz, and Verhagen (2011), Frank (1998), Clementini (2013) and Tenbrink (2011). Levinson's original set of frames (Levinson, 2003) mainly focuses on descriptions captured by *Euclidean coordinate axes*. This includes the construction of coordinate axes on some perceived ground object, and the localisation of figure objects along these axes. While exploring the texts, we quickly realised however that the richness of referencing strategies in mountaineering texts goes well beyond such strategies, exploiting also qualitative (Freksa, 1991), metric or topological relations, several ground objects, as well as metaphorical

strategies including fictive motion (Talmy, 1996). A more comprehensive annotation framework in this respect is *ISO-Space* as proposed by Pustejovsky et
al. (2011), which is based on Spatial-ML (Anderson et al., n.d) and includes
motion events and corresponding paths, as well as mereo-topological relations,
such as 'inside', 'outside', 'overlap' (Herring, Mark, and Egenhofer, 1994). While
this approach acknowledges the relevance of diverse spatial and temporal concepts in spatial referencing, it treats qualitative spatial relations as a superclass
of metric relations, and seems to be restricted to frames having coordinate axes
on a single ground object.

For our purpose, we preferred a less strong spatial commitment. First, we suggest to regard the diversity of referencing strategies on a par, similar to Stock
(2014), and not merely as parameters of the same kind of frame. This means
we treat the different geometric bases of a referencing strategy as independent
from each other. So, for example, using a referencing strategy based on distance
does not necessarily imply any Euclidean axes or even a metric, because the
assumption of Euclidean space are not needed to define a distance or a metric[2].
Furthermore, qualitative relations, such as 'inside', do not necessarily need to be
interpreted as boundary cases of metric relations. And finally, we include the
possibility of a multitude of ground objects. Second, we take seriously the observation that a given spatial referencing strategy, though in itself well defined,
may be expressed in language in diverse and unforeseen ways, forcing us to take
the context of an expression into account (Stock and Yousaf, 2018; Herskovits,
1985)[3]. For example, the preposition 'at' can have different meanings in different contexts (Vasardani et al., 2012). Our intention is therefore to test the quality
of geoparsing rules which can take the context of an expression into account.

Based on the cognitive strategies we encountered in the two mountaineering text sources, we distinguish the following frame categories (cf. Figure 11.2):
*Euclidean frames (EF), Zonal frames (ZF), Topological frames (TF), Linear con-
struction frames (LCF) and Betweenness frames (BF)*. These categories directly
reflect different ways how the corresponding locative expressions could be geo-
metrically translated into the coordinate space of a map[4]:

1. [EF] *Euclidean frames* (EF) cover the well-known types as proposed in
    linguistic literature (Levinson, 2003), see also Scheider et al. (2018) and
    Frank (1998). These frames are used to denote target locations using
    axes in a coordinate system centered on a "ground" object, such that

---

[2] Formally, distances, metric spaces, topologies and Euclidean spaces can all be
    considered independent from each other (Worboys, 1996).
[3] For this very reason, Stock and Yousaf (2018) used a case-based learning approach.
[4] See Stock (2014) for a more comprehensive list of possible strategies. Note that
    many of our strategies can be mapped to this list.

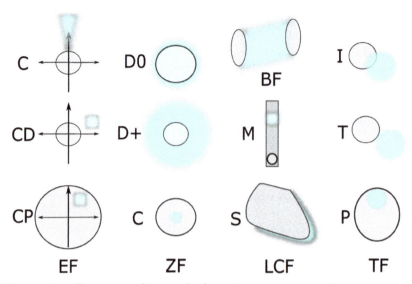

Figure 11.2: Illustration of types of referencing strategies used in environmental narratives. Target locations are indicated in turquoise, ground objects in grey. See text for the meaning of labels.

the main axis is oriented using some perceptual cue, such as a cardinal direction[5]. We only consider *allocentric* frames, where the ground object is some element of the landscape, and not the ego, since ego-centric descriptions did not occur in our example texts. We distinguish the following subtypes: *cardinal direction* (EF C), *cardinal distance* (EF CD), *cardinal part* (EF CP), and *gravitation axis* (see illustrations in Figure 11.2).

"... [(EFC)] [the mountains][(target)] [west of][(orient)] [the moor][(ground)]"

2. [(ZF)] *Zonal frames* (ZF) are used to denote locations purely based on distance measured with respect to a ground object. We distinguish *Zonal frame with zero distance* (ZF D0), *Zonal frame with distance modifier* (ZF D+), and *Zonal frame center* (ZF C).

---

[5] *Absolute* frames are oriented by cardinal directions, while *intrinsic* frames are oriented by the shape of the ground object (such as 'in front'). In environmental narratives, absolute frames seem the dominating orientation strategy, which is why we did not add other sub-types here.

"$(ZFD0)$ At [Rannoch station]$_{(ground)}$ we stepped down from the train ..."

3. (LCF) *Linear construction frames* (*LCF*) are used to denote locations on a one-dimensional path. We distinguish two subtypes depending on how this path is constructed: *Linear construction using shape (LCF S)* and *Linear construction using movement (LCF M)*.

    "... $(LCFS)$ along [the east side of]$_{(orient)}$ [Loch Ba]$_{(ground)}$"

4. (BF) *Betweenness frames* (*BF*) are used to denote places located between two given ground objects. For example, in the following sentence, the location of bog streams is described using two other landscape elements:

    "(BF) Between [the lochs]$_{(ground)}$ and [the peat hags]$_{(ground)}$ [bog streams wriggle]$_{(target)}$"

5. (TF) *Topological frames* (*TF*) are used to denote locations which stand in a mereo-topological relation to ground, such as 'inside', 'outside', 'touch', 'overlap' or 'part of' (Varzi, 1996). Some of these strategies in addition make use of figured features of the ground object in order to further specify the location, which we call *specific part*. We distinguish *Topological frame with touch (TF T) or intersect (TF I)*, *Topological frame with unspecific part (TF P)* and *Topological frame with specific part (TF SP)*.

    'a short and twisted (TFT) [river]$_{(target)}$ linked [Loch Ba]$_{(ground)}$ to [Loch Laidon]$_{(ground)}$'

What we spotted from looking through these examples is that identifying FoR is challenging in particular for the following reasons: (1) The different parameters of the FoR may be distributed across several sentences, making it hard to keep track of them across sentences with NLP. (2) The keywords that may indicate a given type of FoR or parameter can change considerably. For example, a topological relation may be indicated by the word 'link' instead of 'touch', and a vague distance may be given by the word 'well' instead of 'near'. (3) In the case of linear frames, 'fictive motion' (Talmy, 1996) is often required to construct a path in terms of an imagined trail. Thus simulation is required to localise these target locations inside a Geographic Information Systems (GIS). (4) Frames can be nested. Which means that identifying a parameter in parts of speech depends on first identifying another frame, adding considerable complexity to the search task.

11.3   Geoparsing Frames of Reference

In this section, we explain our approach for geoparsing FoRs. We chose a geoparsing framework which addresses word embeddings on different syntactic levels of a sentence, including motion and space keywords.

11.3.1   Perdido parsing rules

The *Perdido Geoparser* annotates different types of information such as named entities (with a special focus on spatial entities), extended named entities (ENE), spatial relations and motion expressions (Gaio and Moncla, 2019). An ENE consists of several overlapping phrase levels where each level is embedded in the previous one. This concept is based on the fact that a proper name can be categorised as pure or descriptive and that the descriptive expansion associated with a name can change the implicit type of the considered entity. The Perdido geoparsing rules have been developed for three romance languages (i.e., French, Spanish and Italian) and they have been used in several research projects to retrieve geographic information and to reconstruct itineraries from texts (Moncla et al., 2019; Moncla et al., 2016). Gaio and Moncla, 2019 argue that for a fine-grained task such as marking, classifying and disambiguating named entities, it is essential to consider the geo-semantic information expressed in the context in order to solve classification and disambiguation issues. Rules implemented in the Perdido Geoparser using cascades of finite-state transducers are a computational synthesis of previous works on how language expresses space and motion (Talmy, 1983; Vandeloise, 1986; Aurnague, 2011). This geoparsing task is based on a bottom-up strategy where each level of embedded entities of the ENE is marked from the pure proper name to the complete expression. Additionally, it can distinguish between two types of spatial entities: 'absolute' referring to standard spatial entities and 'relative' referring to spatial entities associated with spatial relations. More complex expressions involving motion verbs, spatial relations (e.g., spatial prepositions, topological relations, cardinal relations) and spatial ENE are also identified and classified. Examples 9. and 10. show the type of information annotated by Perdido.

9. there is a path that [runs]$_{(motion)}$ [along]$_{(rel)}$ [Loch Laidon]$_{(place)}$ [to]$_{(rel)}$ [Kingshouse]$_{(place)}$ [on]$_{(rel)}$ [the [Glencoe]$_{(place)}$ road]$_{(spatial\ ENE)}$.
10. [From]$_{(rel)}$ [Pitlochry]$_{(place)}$ the route would be either [by]$_{(rel)}$ [Struan]$_{(place)}$ and [Kinloch-Rannoch]$_{(place)}$ or [by]$_{(rel)}$ [Tummelside]$_{(place)}$ [to]$_{(rel)}$ [Tummelbridge]$_{(place)}$ and [straight through to]$_{(rel)}$ [Rannoch]$_{(place)}$.

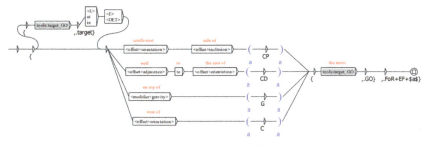

Figure 11.3: Transducer for parsing Euclidean frames.

11.3.2 FoR rules

In this work, we enriched a custom version of the Perdido Geoparser adapted for English texts. Our objective was to transform the existing rules and to add new ones in order to retrieve and classify FoR in environmental narratives. An example of an FoR parsing rule for Euclidean frames implemented using transducers is shown in Figure 11.3, and further ones are described below. A transducer is a local grammar defined as an automaton with an input and output alphabet. It is a type of finite-state machine that makes insertions, replacements and deletions in a text. The cascade of transducers of the Perdido Geoparser is implemented using the CasSys system (Friburger and Maurel, 2004) developed in the Unitex platform[6]. The grammar below mentions the main new transducers that have been added to the Perdido Geoparser for the recognition and classification of FoRs[7]. Transducers are implemented using graphs where each branch refers to a syntactic rule. For instance, for linear construction frames there are two patterns (as shown in the grammar below). Annotations are produced by the output alphabet of the transducers in brackets *{text.,semantic_tag}* with content after the comma referring to the semantic tags that will be transformed into XML elements. This grammar was developed based on a preliminary corpus analysis of the two mountaineering texts (see Section 11.1). The challenge is to build the most exhaustive and precise set of rules. Because of the ambiguity of natural language, however, the same syntactic rule may match different meanings, and may refer to FoRs but also to something else. This distinction cannot be expressed directly in the grammar because it needs external knowledge or context to understand the correct meaning. For this reason the developed grammar is incomplete and highly corpus and task dependent.

---

[6] https://unitexgramlab.org/

[7] For the full list, see our repository https://github.com/simonscheider/FoR.

TF SP = {Target} + modifier topological + GO

TF I = {Target} + modifier intersect + GO + modifier topological + GO

TF T = {Target} + modifier touch + GO + *and* + GO

TF P = {Target} + modifier inclusion + GO

EF CP = {Target + (*at*|*in*)} + modifier orientation + modifier inclusion + GO

EF CD = {Target + (*at*|*in*)} + modifier distance + modifier orientation + GO

EF G = {Target + (*at*|*in*)} + modifier gravity + GO

EF C = {Target + (*at*|*in*)} + modifier orientation + GO

ZF Dn = {Target} + modifier distance + GO

ZF Do = {Target} + modifier location + GO + *and* + GO

ZF C = {Target} + *in* + modifier central + GO

LCF S = {Target} + *along* + modifier orientation + modifier inclusion + GO

LCF M = {Target} + *from* + GO + *to* + GO

BF = {Target} + *between* + GO + *and* + GO

Target = spatial entity

GO = spatial entity + {separator + GO}

Several modifiers were used in our FoR grammar, which consist of lexicons or existing transducers executed at the beginning of the Perdido cascade and then already associated with semantic tags such as <offset+orientation>. These modifiers are expressed by keywords and spatial prepositions in the language. We consider nine types of modifiers: topological (*the mouth of, flows into, …*), intersect (*where, …*), touch (*linked, …*), inclusion (*in, inside, part of, …*), orientation (*north, south, north-east, …*), distance (*near by, away from, close to, …*), gravity (*on top of, …*), central (*middle of, …*), location (*at, where, …*).

11.4   Results and Quality of Geoparsing

In this section, we present and explain our validation results[8]. We start with inter-annotator agreement, and then discuss the quality of automatic parsing on the two mountaineering texts, as well as on an external text source that was not used for exploration and training.

---

[8] All raw resources can be accessed under https://github.com/simonscheider/FoR

11.4.1   Inter-annotator agreement

To assess the comprehensibility of our notion of FoRs, we performed an annotation task on Robert Macfarlane's 'The Wild Places' describing a hike through Rannoch moor (Macfarlane, 2008). For the purpose of this paper, we constrained ourselves to an exemplary approach that has to be expanded in the future, taking into account more texts.

In the vein of the annotation workflow outlined by Pustejovsky and Stubbs (2013), we understand the modeling process as an iterative cycle, in which concepts are tested empirically with the help of manual annotations that in turn serve as a base for a revision of the concepts. To assess the inter-subjectivity of the annotations we let different annotators annotate the same text and calculate their inter-annotator agreement.

For our first annotation round, we formulated guidelines that instructed the annotators to mark up the text with 'FrameOfReference', 'target' and 'groundObject' tags. 'FoR' tags were supposed to be classified according to our categories outlined above. Our first annotation round was performed by four annotators, amongst them one expert annotator from the project team. All annotations were carried out with the webAnno-tool as provided by the CLARIN-D web-service[9].

An error analysis of the first annotation round showed that the syntactic boundaries of the FoRs are very hard to define clearly enough, and thus hard to identify by all annotators, even if the FoR concepts themselves may have a clear definition. Annotators often had different ideas about where exactly a phrase containing a FoR would start or end, and therefore many correct identifications of a given frame type slightly overlapped within a sentence. Therefore, to report on the inter-annotator agreement, we abstain from calculating standard token-based kappa-metrics (Pustejovsky and Stubbs, 2013, pp. 126–134), but rather give the total number of FoRs annotated. Furthermore, we count every annotation that shows an overlapping match as an agreement. In total, the four annotators of the first annotation round classified 40 phrases as FoRs. In 10 cases, all of the annotators agreed, seven phrases were unanimously annotated by three annotators, 11 phrases by two annotators and in 12 cases only one out of the four annotators classified a phrase as a FoR.

To establish a more stable reference for comparison, we revised the expert annotation and defined it as a gold standard (which we later on also used for the comparison of the automatic detection of FoRs in Section 11.4.2). In total, the gold standard features 29 FoRs. Table 11.1 shows the precision and recall of three annotators of the first round compared to the gold standard.

A more in-depth analysis of the errors revealed that a high percentage of disagreement resulted from an insufficient distinction of our notion of FoR to the

---

Annotator 1	Precision	64%
	Recall	79%
Annotator 2	Precision	70%
	Recall	55%
Annotator 3	Precision	59%
	Recall	34%

Table 11.1: Precision and recall of the first annotation round compared to the gold standard.

Annotator 4	Precision	64%
	Recall	62%
Annotator 5	Precision	65%
	Recall	72%
Annotator 6	Precision	56%
	Recall	79%

Table 11.2: Precision and recall of the second annotation round compared to the gold standard.

concept of motion: For our model, we want to exclude expressions that indicate a motion without referring to a place, whereas often annotators tended to annotate motion phrases as linear construction frames. Therefore, we revised our *guidelines*[10] on this behalf and explicitly told annotators not to tag motion expressions.

On the basis of these guidelines, we performed a second annotation round with three more annotators. However, even though the quality slightly improved in this round, the annotators still showed a significant amount of disagreement. In total, 52 phrases were classified as FoRs. In 15 cases all four annotators (including the gold standard annotation) agreed, in six cases three out of the four, in 12 cases only two and in 19 cases annotators stayed on their own in their decision to annotate a phrase as FoR. Table 11.2 shows precision and recall in comparison to the gold standard.

The rather low agreement shows that FoRs are a difficult concept that still needs clarification. Furthermore, it is apparent that more training is needed for interpreting FoRs in texts in a consistent way. A further refinement of the annotation guidelines remains a task for future work.

---

[10] Available under http://geographicknowledge.de/pdf/AnnGuiEnv.html.

11.4.2   Quality on mountaineering texts

For our parser experiments, we processed both McFarlane's 'The Wild Places' and Murray's 'Undiscovered Scotland' with the proposed FoR parsing rules implemented in the custom version of the Perdido Geoparser (as described in Section 11.3.2). In order to assess the quality of the automatic annotation, we compare the results with the gold standard annotation (see Section 11.4.1). For this purpose, we use three different metrics: precision, recall and Slot Error Rate (SER)(Makhoul et al., 1999). SER takes into account different types of errors: substitutions (S), insertions (I) and deletions (D). Substitution errors are of three kinds: wrong boundaries identification (B), wrong classification (T) and both (CT). Insertion errors refer to false positives (i.e., entities identified by the system that do not exist in the gold standard) and deletions errors refer to false negatives (i.e., entities existing in the gold standard that are not identified by the system). In addition to the precision and recall measures, the SER allows us to consider not only the identification of the expressions but also their classification.

Our gold standard for evaluation is composed of 69 FoR expressions, 76 ground objects and 10 target entities. Table 11.3 shows the number of each error type and the scores for different evaluation measures over the MacFarlane and Murray chapters. We notice that the distribution of all types of FoR expression is not homogeneous in our gold standard. For instance, we only have two betweenness FoR but 10 Euclidean or 22 zonal FoR. This implies that betweenness FoR results will not be representative and meaningful. However, one interesting observation is that all types of FoR have a rather high precision score.

Another interesting result is that the greatest number of errors refers to deletions (i.e., false negative) and implies a rather bad recall score (37,68%). This means that our system missed a lot of FoRs (44 over 69 for FoR expressions, 46 over 76 for 'Ground object' and eight over 10 for 'Target entities'). This can be explained by the fact that we built the rules based on a preliminary analysis of a very limited corpus. We thus will need many more examples in order to build more exhaustive rules. This also shows there might be a potential for machine learning-based approaches, however, this will also need a larger manually annotated dataset in order to train a model.

11.4.3   Precision on adventure novel

In order to measure the quality of automatic geoparsing on an external source, we ran Perdido over the first four chapters of the novel *Kidnapped* by R. L. Stevenson. These chapters describe David Belfours journey to his uncle's house in Scotland. Within the first 3311 sentences, Perdido found 80 occurrences of a reference frame, that is, one occurrence per 41 sentences. We went through all of these text snippets and manually evaluated their correctness, in order to measure the *precision*.

As a result, 67 of these 80 occurrences were correct, which is a precision of **84%**. The stacked bar chart in Figure 11.4 shows the distribution over the different types of frames, where grey bars indicate the wrongly annotated cases. From this diagram, we can see that zonal frames were detected most often, followed by topological and Euclidean frames. Betweenness frames and linear construction frames are most seldom. Regarding the precision per type, we can see that it differs largely between the different types of frames. Euclidean, betweenness and linear construction frames were detected without any false positives, and only three out of 32 annotations of topological frames were incorrect. This corresponds to a precision of **90%**. The precision of zonal frames is worse, but still **70%**. It seems thus our rules work rather well also on external text sources for the considered frame types. We expect however that recall, which we did not test on the novel, should be equally worse as in the training corpus. Also, note that zonal frames are not only the most challenging case but also the most frequently occurring type.

To spot the syntactic reasons for this pattern, we plotted the frequency of different modifiers and keywords (every word except the ground and target objects) within the annotated text snippets, for all true-positive as well as

	FoR						Ground objects	Target entities
	all	btw	eucl	linear	topo	zonal		
Gold standard	69	2	10	18	17	22	76	10
Correct	18	1	4	4	2	7	28	1
(B)	6	1	1	0	2	2	2	1
(D)	44	0	5	14	12	13	46	8
(I)	7	0	3	0	0	4	10	3
(T)	1	0	0	0	0	1	0	0
(CT)	1	0	0	0	1	0	0	0
SER	80,43%	25%	85%	77,78%	82,35%	84,09%	75%	115%
Recall	37,68%	100%	50%	22,22%	29,41%	45,45%	39.47%	20%
Precision	78,79%	100%	62,5%	100%	100%	71,43%	75%	40%
Classification precision	72,73%	100%	62,5%	100%	80%	64,29%	75%	40%
Boundaries precision	57,58%	50%	50%	100%	40%	57,14%	70%	20%

Table 11.3: Number of errors and evaluation scores for our gold standard corpus.

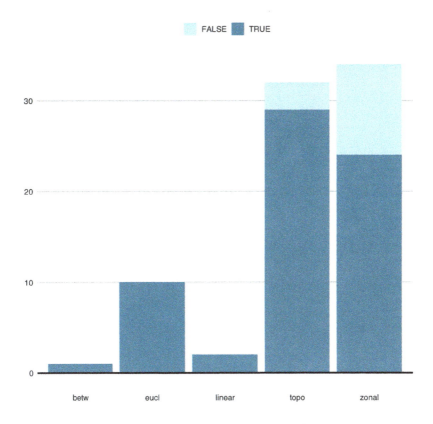

Figure 11.4: Precision of detecting different types of reference frames in the first 3311 sentences of the 1886 novel *Kidnapped*, including ZF ('zonal'), TF ('topo'), LCF ('linear'), EF ('eucl') and BF ('betw') frames.

false-positive annotations. Figure 11.5 shows a wordcloud for each type of frame over true- and false-positives. It can be seen there is no spottable difference in keywords between true and false zonal frames (Figure 11.5d), where locative prepositions 'where' and 'at' are used in both cases. An example for an erroneous annotation is:

'Looking $^{(ZF)}$ at [the shore]$_{(ground)}$ ...'

where the preposition 'at' is not locative, but used instead for indicating the direction of view. The topological frame errors (Figure 11.5b) have mostly to do with the motion indicator 'through', whereas a large diversity of expressions is correctly exploited in both the topological (Figure 11.5a) and the Euclidean case (Figure 11.5e).

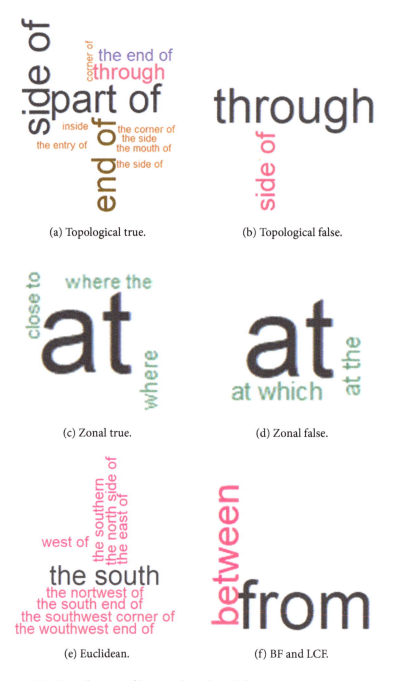

(a) Topological true.

(b) Topological false.

(c) Zonal true.

(d) Zonal false.

(e) Euclidean.

(f) BF and LCF.

Figure 11.5: Distribution of keywords and modifiers over true- and false-positive frame annotations.

11.5   Discussion and Future Work

Though our results are based on a limited corpus and thus are preliminary, we think they offer three main insights:

1. The suggested FoR typology seems to cover many relevant strategies in the chosen texts, however, it is almost certainly incomplete. We therefore expect that the diversity of cognitive spatial referencing strategies is far from exhausted by the suggested FoR model. Research that tests or extends this model is therefore needed. Furthermore, we believe that cognitive research (beyond text analysis) is needed in order to better understand which cognitive referencing strategy is actually used by readers when interpreting a text. The fact that a particular keyword is used in a particular context is only a very vague hint at the spatial cognitive strategy, and so text alone cannot be decisive. For this reason, empirical research combining spatial cognition and linguistics should focus on experiments that actually demonstrate and highlight the strategy hidden in the syntactic depths of environmental texts (Montello, 2009).

2. Even in case we might come up with clear, empirically validated models of FoR concepts, our study demonstrates that this does not yet mean clarity and ease of annotation for human annotators. As the rather low rate of inter-annotator agreement on types of FoR illustrates, human annotators frequently confuse actual movements with fictive motions and linear path references to space, as well as the strategies implied by mereology, topology, direction and distance in space. Another difficulty concerns the inherent vagueness of deciding which parts of speech should belong to a given frame (boundary errors). It seems that deciding on the precise sequence of words which denote a frame is hard, even if the frame used in a sentence might be easy to spot. This is true, by the way, for both human annotators and geoparsers. Annotators therefore need to be trained experts in order to be considered trustworthy producers of gold standards for information retrieval.

3. It is thus surprising that under these circumstances, the classification precision of a geoparser using transducer-based rules designed on a small training sample is considerably high ($\sim$ 70–80%), both on training and test data (when disregarding boundary errors). The fact that recall, on the other hand, is so low ($\sim$ 40%) shows that the main challenge of automatically discovering FoRs is not finding a robust model for a particular strategy, but rather handling the diversity of frames in their diverse natural language forms. In addition, we have a severe *cold-start problem*: In order to exploit machine learning to tackle diversity, we would need a large ground truth data set, which is lacking precisely for the reasons mentioned under point 2. It is thus an

important future task to establish a gold standard corpus of annotated FoRs of a sufficient size and variability.

What do these results tell us about the possibilities of automated spatial referencing and alignment of narrative texts? It seems that hand-crafted grammatical rules, as illustrated in this study, can in fact be used to reliably extract indirect spatial references, albeit only for a limited set of linguistic strategies. Once discovered, strategies allow approximating unnamed places of the described journey, for example, starting from a place in the middle of the moor, towards a place next to a Loch, and from there to a place near a certain mountain. Mapping of journeys in this way would allow us to find out to what extent the trails described by Murray, McFarlane and Stevenson really overlapped, and whether these authors saw the landscape from comparable vantage points. To this end, future work should investigate geometric approximations of any given FoR reference.

11.6   Conclusion

In this chapter, we have investigated the possibility of discovering indirect spatial references in environmental narratives, in order to align these narratives with the landscape features they describe. Based on two mountaineering texts over Scotland, we have explored the referencing strategies behind expressions used to localise the narrative within a landscape, and we have suggested a FoR typology on this basis, treating different geometric referencing strategies on a par. We have then designed rules and trained a transducer-based geoparser to automatically find these frames, before applying it to an external text for testing purposes. We compared the results with a manually annotated gold standard, which was also tested for inter-annotator agreement. Our results show that on the one hand, manual annotation of FoR types is surprisingly hard and annotators are in disagreement. On the other hand, classifier precision of the geoparser is considerably high on both the training and test data. The biggest challenge seems to be the low recall rate, which underlines our principle insight that discovering FoRs is primarily a problem of coping with the semantic and syntactic diversity of spatial referencing in texts. Yet, due to the considerably high precision, we believe automated parsing of indirect spatial references is possible and could be used to align documents and vantage points based on equivalent places described in these texts.

References

Anderson, Dave, Jade Goldstein-Stewart, Amal Fayad-Beidas, Dave Harris, Dulip Herath, Qian Hu, Janet Hitzeman, Seok Bae Jang, Inderjeet Mani, James Pustejovsky, et al. (n.d). *SpatialML: Annotation scheme for marking spatial expressions in natural language*. Tech. rep. Bedford.

Aurnague, Michel (2011). "How motion verbs are spatial: The spatial foundations of intransitive motion verbs in French". In: *Lingvisticae Investigationes* 34.1, pp. 1–34. DOI: 10.1075/li.34.1.01aur.

Burenhult, Niclas and Stephen C Levinson (2008). "Language and landscape: A cross-linguistic perspective". In: *Language Sciences* 30.2-3, pp. 135–150. DOI: 10.1016/j.langsci.2006.12.028.

Chen, Hao, Maria Vasardani, and Stephan Winter (2018). "Georeferencing places from collective human descriptions using place graphs". In: *J. Spatial Information Science* 17.1, pp. 31–62. DOI: 10.5311/JOSIS.2018.17.417.

Clementini, Eliseo (2013). "Directional relations and frames of reference". In: *GeoInformatica* 17.2, pp. 235–255. DOI: 10.1007/s10707-011-0147-2.

Clementini, Eliseo, Paolino Di Felice, and Daniel Hernández (1997). "Qualitative representation of positional information". In: *Artificial intelligence* 95.2, pp. 317–356. DOI: 10.1016/S0004-3702(97)00046-5.

Frank, Andrew U (1998). "Formal models for cognition–taxonomy of spatial location description and frames of reference". In: *Spatial cognition*. Berlin/Heidelberg: Springer, pp. 293–312. DOI: 10.1007/3-540-69342-4_14.

Freksa, Christian (1991). "Qualitative spatial reasoning". In: *Cognitive and linguistic aspects of geographic space*. Berlin/Heidelberg: Springer, pp. 361–372. DOI: 10.1007/978-94-011-2606-9_20.

Friburger, Nathalie and Denis Maurel (2004). "Finite-state transducer cascades to extract named entities in texts". In: *Theoretical Computer Science*. Implementation and Application of Automata 313.1, pp. 93–104. ISSN: 0304-3975. DOI: 10.1016/j.tcs.2003.10.007.

Gaio, Mauro and Ludovic Moncla (2019). "Geoparsing and geocoding places in a dynamic space context. The case of hiking descriptions". In: *The Semantics of Dynamic Space in French. Descriptive, experimental and formal studies on motion expression*. Ed. by Michel Aurnague and Dejan Stosic. Vol. 66. Human Cognitive Processing. Amsterdam: John Benjamins. Chap. 10, pp. 354–386. DOI: 10.1075/hcp.66.10gai.

Herring, John, David M Mark, and Max J Egenhofer (1994). *The 9-intersection formalism and its use for natural-language spatial predicates*. Tech. rep. Bedford.

Herskovits, Annette (1985). "Semantics and pragmatics of locative expressions". In: *Cognitive Science* 9.3, pp. 341–378. DOI: 10.1207/s15516709cog0903_3.

Levinson, Stephen C (2003). *Space in language and cognition: Explorations in cognitive diversity*. Vol. 5. Cambridge: Cambridge University Press. DOI: 10.1017/CBO9780511613609.

Macfarlane, Robert (2008). *The wild places*. London: Granta Books.

Makhoul, John, Francis Kubala, Richard Schwartz, Ralph Weischedel, et al. (1999). "Performance measures for information extraction". In: *Proceedings of DARPA broadcast news workshop*. Herndon, VA, pp. 249–252.

Mark, David M, Christian Freksa, Stephen C Hirtle, Robert Lloyd, and Barbara Tversky (1999). "Cognitive models of geographical space". In: *International Journal of Geographical Information Science* 13.8, pp. 747–774. DOI: 10.1080/136588199241003.

Moncla, Ludovic, Walter Renteria-Agualimpia, Javier Nogueras-Iso, and Mauro Gaio (2014). "Geocoding for texts with fine-grain toponyms: An experiment on a geoparsed hiking descriptions corpus". In: *Proceedings of the 22nd acm sigspatial international conference on advances in geographic information systems*. New York: ACM, pp. 183–192. DOI: 10.1145/2666310.2666386.

Moncla, Ludovic, Mauro Gaio, Javier Nogueras-Iso, and Sébastien Mustière (2016). "Reconstruction of itineraries from annotated text with an informed spanning tree algorithm". In: *International Journal of Geographical Information Science* 30.6, pp. 1137–1160. DOI: 10.1080/13658816.2015.1108422.

Moncla, Ludovic, Mauro Gaio, Thierry Joliveau, Yves-François Le Lay, Noémie Boeglin, and Pierre-Olivier Mazagol (2019). "Mapping urban fingerprints of odonyms automatically extracted from French novels". In: *International Journal of Geographical Information Science* 33.12, pp. 2477–2497. DOI: 10. 1080/13658816.2019.1584804.

Montello, Daniel R (2009). "Cognitive research in GIScience: Recent achievements and future prospects". In: *Geography Compass* 3.5, pp. 1824–1840. DOI: 10.1111/j.1749-8198.2009.00273.x.

Murray, W.H. (2003). *Undiscovered Scotland*. Baton Wicks Publications, pp. 155–163.

Palmer, Bill (2002). "Absolute spatial reference and the grammaticalisation of perceptually salient phenomena". In: *Representing space in Oceania: Culture in language and mind*, pp. 107–157.

Purves, Ross S., Paul Clough, Christopher B. Jones, Mark H. Hall, and Vanessa Murdock (2018). "Geographic information retrieval: Progress and challenges in spatial search of text". In: *Foundations and Trends in Information Retrieval* 12.2-3, pp. 164–318. DOI: 10.1561/1500000034.

Pustejovsky, James, Jessica L Moszkowicz, and Marc Verhagen (2011). "Using ISO-Space for annotating spatial information". In: *Proceedings of the International Conference on Spatial Information Theory*.

Pustejovsky, James and Amber Stubbs (2013). *Natural language annotation for machine learning*. Sebastopol: O'Reilly, pp. 155–163.

Scheider, Simon and Ross Purves (2013). "Semantic place localization from narratives." In: *COMP@ SIGSPATIAL*, pp. 16–19.

Scheider, Simon, Jürgen Hahn, Paul Weiser, and Werner Kuhn (2018). "Computing with cognitive spatial frames of reference in GIS". In: *Transactions in GIS* 22.5, pp. 1083–1104. DOI: 10.1111/tgis.12318.

Stock, Kristin (2014). "A geometric configuration ontology to support spatial querying". In: *Connecting a Digital Europe Through Location and Place. The 17th AGILE International Conference on Geographic Information Science.*

Stock, Kristin and Javid Yousaf (2018). "Context-aware automated interpretation of elaborate natural language descriptions of location through learning from empirical data". In: *International Journal of Geographical Information Science* 32.6, pp. 1087–1116. DOI: 10.1080/13658816.2018.1432861.

Talmy, Leonard (1983). *How language structures space.* anglais. Berkeley Cognitive Science Report 4. Berkeley, CA, Etats-Unis: Cognitive Science Program, Institute of Cognitive Studies, University of California at Berkeley.

— (1996). "Language and space". In: vol. 21. Cambridge, Mass. London: MIT Press. Chap. Fictive motion in language and "ception", pp. 211–276.

Tenbrink, Thora (2011). "Reference frames of space and time in language". In: *Journal of Pragmatics* 43.3, pp. 704–722. DOI: 10.1016/j.pragma.2010.06.020.

Tuan, Yi-Fu (1991). "Language and the making of place: A narrative-descriptive approach". In: *Annals of the Association of American Geographers* 81.4, pp. 684–696. DOI: 10.1111/j.1467-8306.1991.tb01715.x.

Vandeloise, Claude (1986). *L'Espace en français. Sémantique des prépositions spatiales.* Paris: Editions du Seuil.

Varzi, Achille C (1996). "Parts, wholes, and part-whole relations: The prospects of mereotopology". In: *Data & Knowledge Engineering* 20.3, pp. 259–286. DOI: 10.1016/S0169-023X(96)00017-1.

Vasardani, Maria, Stephan Winter, Kai-Florian Richter, Lesley Stirling, and Daniela Richter (2012). "Spatial interpretations of preposition at". In: *Proceedings of the 1st ACM SIGSPATIAL International Workshop on Crowdsourced and Volunteered Geographic Information.* ACM, pp. 46–53. DOI: 10. 1145/2442952.2442961.

Viehhauser, Gabriel and Florian Barth (2017). "Towards a digital narratology of space". In: *Digital Humanities 2017 Conference Abstracts* (McGill University & Université de Montréal Montreal Canada), pp. 643–646.

Worboys, Michael F (1996). "Metrics and topologies for geographic space". In: *Advances in Geographic Information Systems Research II: Proceedings of the International Symposium on Spatial Data Handling, Delft, pages A.* Vol. 7.

CHAPTER 12

Surveying the Terrain and Looking Forward

Ross S. Purves

Department of Geography; URPP Language and Space, University of Zurich,
Switzerland

Olga Koblet

Department of Geography, University of Zurich, Switzerland

Benjamin Adams

Department of Computer Science and Software Engineering, University of
Canterbury, Christchurch, New Zealand

In our introduction we described the three pillars of the workshop which gave
rise to this book. The first was concerned with the identification of themes
related to the environment, and the nature of the multidisciplinary questions which might be explored through text. The second concentrated on the
resources and methods available to us which might enable addressing these
questions, and the third focused on the development of individual, illustrative
case studies. These pillars give a useful framework for some concluding remarks,
identifying areas of common ground and potential for future work. These
remarks are structured around four elements. Firstly, we discuss the nature of

How to cite this book chapter:
Purves, Ross S., Olga Koblet, and Benjamin Adams (2022). "Surveying the
Terrain and Looking Forward." In: *Unlocking Environmental Narratives: Towards
Understanding Human Environment Interactions through Computational Text
Analysis.* Ed. by Ross S. Purves, Olga Koblet, and Benjamin Adams.
London: Ubiquity Press, pp. 233–244. DOI: https://doi.org/10.5334/bcs.l.
License: CC-BY 4.0

the collections analysed, before moving to the methods used to explore these data. We then step back, and explore not only the nature of the questions posed in the book, but discuss the potential and limitations of some of the results presented in the book. Finally, we set out a research agenda for future work, picking out some potential themes for research. We do not make claims for exhaustivity in any of these elements – rather, our aim is to illustrate the, largely to date unrealised, potential for the analysis of unstructured text in addressing pressing scientific, societal and policy-related environmental research questions.

Our case studies embraced a very broad range of resources for their analysis. The nature of these sources reflected not only the types of questions being explored, but also the research interests and backgrounds of those exploring them. Thus, for example, Tobias Zuerrer's starting point given his background in English literature, was a classic novel of the 19th century, Mary Shelley's *Frankenstein*. As an out-of-copyright classic novel, where the protagonists travel in a range of landscapes, it provided an accessible and appropriate starting point for questions concerning ways in which the industrial revolution was reflected in descriptions of urban and natural scenes.

Sarah Luria and Ricardo Campos were concerned with similar questions, but chose a particular location, the historical Canal District of Worchester, Massachusetts and diverse narratives about the deindustrialisation of the city over time. By definition, their study required a collection of texts from different times and different authors. Since one author, Sarah, was familiar with the story of the Canal District, she hand-built a small but very diverse corpus of texts capturing these different voices in quite different genres, ranging from poetry through to news reporting. The selection of texts was neither objective nor exhaustive, but (in common with many approaches from the humanities) no such claims are made. Karen Jones, Diana Maynard and Flurina Wartmann took a somewhat similar approach to building a corpus of historical documents about Loch Lomond in Scotland. However, their approach was different in that they started from a small set of historical documents identified by searching online archives, before comparing their documents with lists compiled about the region. Their aim was therefore to create a more systematic corpus of all travel writing about Loch Lomond in the 18th and 19th centuries, and they acknowledge that doing so would likely require digitisation of sources currently only available in analogue form. Unlike Sarah and Ricardo, Karen and her colleagues focused on a particular genre of writing, that of travel guides, implying that these sources are likely to throw light on views of the region from a particular perspective, which may be quite different to that experienced by its inhabitants. Nonetheless, both of these studies used geographically constrained collections to specifically ask questions about particular locations.

Two further studies also chose to focus their research on a particular location, though in these cases the choice of place, Rannoch Moor, was an artefact of a set of texts used in discussions at the workshop from which this book stemmed. Joanna Taylor and Ben Adams used writing from W. H. Murray and Robert Macfarlane about Rannoch Moor as a starting point for a discussion about the influence of gender on writing about wild places, adding travel writing and poems to build an initial corpus. However, the location is not central to their hypotheses, and they extended their work to include more nature writing, in the form of the Guardian's Country Diary column. Here, Joanna and Ben deliberately choose a collection of texts of the same genre and from the same source available online through an application programming interface (API). Their second collection was much larger than the small collections used by our authors so far, consisting of more than 6000 articles. Simon Scheider, Ludovic Moncla and Gabriel Viehhauser started by analysing the same two pieces from Murray and Macfarlane, but with a very different aim. They wished to explore how frames of reference are used in writing, in order to improve methods which extract locations from such texts. They used these texts to develop a model, which they then tested on a different genre of text, the historical novel *Kidnapped* by Robert Louis Stevenson. This choice again gave the team access to an out-of-copyright and digitised source, which much like *Frankenstein* included rich spatial texts describing the main protagonists' journeys through Scotland.

The final two case studies were less directly concerned with explicit spatial locations, and more so with writing about specific subjects. Joanna Taylor and her colleagues investigated how landscape is assigned value in a very technical genre of writing, so-called Landscape Character Assessments, which are widely used in landscape policy and management in the UK. This technical form of writing meant that the authors could look for specific elements of documents in their corpus, but they were hindered by the structure of the documents available online. Unlike most of the other texts analysed in the book, these were often Portable Document Formats (PDFs), meaning that before applying computational methods to raw text this text first had to be extracted from structured documents, and elements such as figure captions, text contained in tables, information boxes had to be filtered from the main body of the text. This case study illustrates clearly that even where a specific genre is digitally available, these texts may require considerable preprocessing before substantive analysis can commence. It also hints at the related issue of context – how much, for example, of the writing in the historical guidebooks analysed by Karen Jones and colleagues, rely on context given by accompanying maps and sketches, and what is lost when our analysis ignores this? Katrín Lund, Ludovic Moncla and Gabriel Viehhauser took a different starting point, a specific location in Iceland, for their diachronic study of how glaciers are captured in narrative. In contrast to the other works we report on, they started with much larger initial collections in multiple languages and of quite different genres. These included British parliamentary proceedings, a well-known German news magazine, *Der Spiegel*, and German versions of the Swiss Alpine Club's yearbook from the corpus Text+Berg. They extracted potentially relevant articles using keyword searches for 'glacier' and its synonyms. However, as they acknowledge, such an approach to corpus building does not deal with word sense ambiguity, for example with respect to metaphorical uses of glacier in text. Since such usage can be both a function of genre and time, it is an important consideration in not only the building of corpora but also the filtering of relevant texts for further analysis.

To scholars from the humanities, this discussion of the nature of sources and their origins is perhaps obvious. However, for those more accustomed to working with other forms of digital data – for example, in the form of terrain models or land cover data – it is important to emphasise the challenges in understanding how the collections selected can, and do, influence our analysis. Equally important is a recognition that large collections are not necessarily more effective ways of studying specific questions computationally, and that the importance of domain knowledge with respect to the theme under investigation cannot be overestimated. Our case studies illustrate the breadth of sources amenable to computational analysis, and leave us ready to discuss the methods applied by our teams. However, before we do so it is important at this point to make some caveats. In particular, most of the texts we worked with were in English, and their settings were European and North American. Thinking about ways of including both questions and sources from the Global South is an important challenge, and one that we do not address here.

This book is predicated on the potential of computational analysis of environmental narratives to extract information not otherwise accessible. Each of our teams used a combination of techniques to analyse text, and importantly for the reader of this book, the focus was very much on the use of existing approaches, rather than development of new methods. Many of these methods were similar to those we introduced in Chapter 3, and together they give a good overview of starting points for future work.

Our first case study, from Katrín Lund and colleagues sought to explore the narratives existing in three distinctive collections over time. Computationally, standard methods were applied to characterise the three corpora over time, using simple frequency-based approaches to suggest potential themes in these corpora. Such methods are essentially language-independent, and the authors chose to remove only stop words before analysing the remaining tokens in each corpus. By using collocation, the team zoomed in on some of the typical themes discussed with respect to glaciers, though arguably those identified are relatively unsurprising given the nature of the corpora. Through microreading of individual texts the importance of the use of 'glacier' metaphorically, particularly in the parliamentary corpus, became apparent. This importance of metaphor, and the potential of its prevalence as a function of individual collections is an important methodological consideration, since its detection requires examination of the source material in detail.

The computational approach taken by Sarah Luria and Ricardo Campos was in many ways similar to that of Katrín, Ludovic and Gabriel. Just like the previous study, the aim was to identify salient terms used in documents, however rather than characterising an entire corpus, Sarah and Ricardo summarised individual documents from a hand-picked corpus. Because some of the documents in this corpus were not in a suitable digital form, an initial pre-processing step using optical character recognition (OCR) was required and, as is often the case with historical texts, some post-processing was also necessary to deal with errors in the OCR process. Rather than simply extracting high-frequency terms or significant collocates, Sarah and Ricardo applied a bespoke piece of software, Yake!. Yake! uses an unsupervised approach to extract keywords and is language-independent. This means it can be applied directly to individual documents or a complete corpus. In practice, simple features such as frequency and collocates are used to identify important keywords, which may also take the form of n-grams. Because Yake! was used on a small corpus, and because Sarah had detailed knowledge about both the process and the sources being explored, it was possible to perform a much more detailed, but qualitative, evaluation of the terms extracted and represented as word clouds. Interpreting these word clouds was only possible given Sarah's underlying knowledge of the corpus, and the conclusions drawn are thus inherently dependent on both the macroreading performed by Yake! and Sarah's microreading of the individual texts.

Karen Jones, Diana Maynard and Flurina Wartmann investigated historical travel writing about Loch Lomond in Scotland in their case study. Similar to Sarah and Ricardo, after selecting an initial collection of documents, preprocessing was necessary, though in this case with the aim of removing extraneous material such as indexes and dividing books into small enough sections for processing. Like the previous two case studies, Karen and colleagues looked at individual words and their use in their collections, however their starting point is more semantically constrained. Using the text analysis toolkit GATE, which Diana has played a key role in developing, the team identified parts of speech and extracted landscape elements and place names. Doing so required the creation of curated lists of relevant terms. Unlike the first two case studies, Karen and her colleagues explicitly link narrative to space by, for example, mapping the order in which locations were discussed in the texts, and compared these computationally extracted lists with microreadings of the text.

They also explored co-occurrences, but did so for their lists of landscape terms and investigated their relationship with a list of more abstract terms relating to landscape. Like Sarah and Ricardo, the importance of interpreting and discussing the computational results from a particular perspective, in this case that of an environmental historian is once again a key element of the case study.

Tobias Zuerrer also investigated landscape perception in text, setting out to explore a potential dichotomy (that of urban and natural landscapes) using a curated set of seed terms containing both generic landscape terms and toponyms. Tobias used an off-the-shelf tool, AntConc, to explore the frequency of seed terms, and used concordances to perform a simple microreading and annotate positive and negative connotations. He very effectively demonstrates how an existing tool can be used to analyse texts with no need for programming skills, and provides an excellent example of what is achievable through carefully considered questions and existing tools with respect to the computational analysis of text.

The first four case studies all took essentially exploratory approaches, using simple methods such as term frequency, collocation and order to select and discuss particular environmental narratives, visualising these through word clouds, tables and simple maps. Joanna Taylor and Ben Adams took a similar initial starting point to exploring how gender influenced the use of pronouns in writing about Rannoch Moor, using concordance plots and collocates. Having demonstrated that authors, in an initially small corpus, appeared to describe the landscape differently according to gender, Joanna and Ben tested their hypothesis by developing a supervised classifier capable of assigning gender to a text based on the language used. Doing so required training and test data with gender annotations, and they used an existing tool to automatically assign annotation based on forenames. For the classification itself, they used a well-known classifier, Naive Bayes, which treats a document as a bag of words and assigns it a probability of belonging to a particular class. In a back and forth that lies at the core of the methodological approach taken, Joanna and Ben then identified collocates used with pronouns in their corpus and used these as a basis for further microreading.

Joanna and Ben relied on a supervised classification in their study of diversity of voices about wild places. Such a classification implied in turn that classes exist. In their investigation of value in the widely used instrument Landscape Character Assessment, Joanna Taylor, Meladel Mistica, Graham Fairclough and Timothy Baldwin took a different approach and used an unsupervised approach, topic modelling. Similar to the work of Karen and colleagues, an important methodological pre-processing step was the extraction of relevant text. Here, the challenge was not noisy Optical Character Recognition (OCR) on historical texts, but rather extracting meaningful structure from rich PDF documents which use text boxes, figures, and tables to ease reading and salience for humans. After preparing the texts, topic modeling, and specifically latent Dirichlet allocation (LDA), assigned individual words the probabilities of being associated with particular topics. Since topics also consist of statistically related words, Joanna and her colleagues used LDA as a way of exploring the extent to which value was explicitly and implicitly described in LCA. Interestingly, using computational methods allowed the authors to reframe their understanding of value in terms of LCA through microreading. Here, somewhat in contrast to the earlier studies, the computational analysis of the texts explicitly suggested new ways of interpreting the content, through a change in the way in which

LCAs were read paying attention to properties of the text revealed through topic modelling.

The final case study, from Simon Scheider, Ludovic Moncla and Gabriel Viehhauser, explored how space was referenced in environmental narratives. Like several other studies, the authors annotated texts, in this case though as a first step in defining differing ways in which frames of references were constructed. Using this annotation, it was possible to propose a set of rules which were then implemented in the Perdido Geoparser. Thus, like all of the other studies, Simon and colleagues applied well-tested existing methods to their problem. They emphasised the challenges in annotating frames of reference consistently, and the need for an iterative process to capture a concept about which humans do not always agree. The results of applying their approach to a new text illustrated that the challenge is not simply encoding rules correctly (as captured by precision), but having a training data set with sufficient examples to cover possible cases – as reflected by the low recall of their approach. Crucially, the methods developed by Simon, Ludovic and Gabriel allow us to start to extract complex spatial frames of reference, such as are commonly found in the texts which form the subject matter of this book.

In listing the methodological approaches taken in this book, a few points stand out. Firstly, and most importantly, all of our authors adopted what Joanna Taylor called a 'multiscalar analysis' – that is to say used different approaches to collecting, analysing and interpreting texts, moving fluidly back and forth between macro and microreadings. This multiscalar approach was enabled in all but one single-authored piece by inherently multidisciplinary teams, who worked together to bring a range of approaches to the table. This is perhaps best illustrated by Sarah and Ricardo's piece, which by documenting some initial misunderstandings helps uncover the need for a constant dialogue in such work.

Secondly, all of our teams made use of relatively long established methods, rather than state of the art machine learning approaches which are currently being applied to a wide range of tasks in natural language processing and are gaining popularity in the digital humanities. This does not, we think, mean that these methods do not have potential in the analysis of environmental narratives. Rather, however, where simple off-the-shelf methods allow first exploratory insights, these may be an effective way of starting discussions between disciplines such as those exemplified in our case studies.

As we argued in the introduction, starting these discussions also requires that meaningful questions are identified. A strength, we would argue, of the case studies here is the interdisciplinary inputs to both the research questions and the methodological approaches taken. Arguably, the results are often modest, and many are either inconclusive or suggest starting points for further work rather than delivering deep insights. We have already shown great variation in the nature of the collections analysed which contrasted strongly with the relatively consistent use of standard off-the-shelf methods in much of the analysis. What though of the questions asked in our studies – do these reflect the breadth we found in the ways that workshop attendees asked questions when given a single short text to reflect on, or does the relative homogeneity of the methods applied constrain the ways in which these texts are approached computationally?

In the introduction we used a simple framework – the 5Ws & H (what, why, when, where, who and how) as a tool to categorise ways in which questions were asked of our texts. What happens when we do the same with our case studies?

Katrín and colleagues set out to explore the influence of different voices in narratives about glaciers at multiple locations and times. Computationally, they used corpora in different languages (English and German), of different genres (parliamentary records, news reporting and mountaineering yearbooks) and with historical depth as proxies for the questions where, who and when? In practice, they could start to 'track down the traces of the multitude of voices ... hidden in large text corpora', but also struggled to reconcile the qualitative nature of the questions suggested by microreading of environmental narratives with the broad conclusions that could be drawn through a macroreading limited by the need to deal more effectively with metaphor.

These limitations are interesting, since in many ways the approach taken by Sarah and Ricardo was very similar. They too wanted to explore narratives about a particular location – the Canal District (where), changes in these narratives over time (when), and explore both the voices (who) and forms (how) of these narratives. Their focus on a particular district, their use of a curated corpus, whose constituent parts one of the authors was very familiar with, and their hermeneutic back and forth led them to be very positive about the possibilities of understanding the process of revitalisation and gentrification in the Canal District through text analysis. We believe this points to an important dichotomy between the needs of scholars concerned with environmental narratives and those interested in computational methods. Despite the allure of running methods over very large corpora, these methods are essentially limited by the need for microreading to understand context, allowing an iterative back and forth with the material being researched.

This in turn leads us to an argument for what Joanna Taylor and colleagues argued for as multiscalar approaches. In their approach to exploring a very specific genre of document (landscape character assessments), the team focused on understanding a specific aspect – value (what?) and the ways it was implicitly and explicitly described in these documents (how?). By again building in a back and forth between computational analysis and microreadings of their texts, Joanna and team were able to point to some discrepancies in the ways LCA sets out to describe character without ascribing value, and the reality of the close link between character and distinctiveness and value. Furthermore, by identifying a potential link in who wrote these LCAs, and in particular their geographic origins further potential questions are hinted at (where, and by whom is landscape value ascribed?).

Perhaps unsurprisingly Joanna, this time with Ben Adams, once again took a multiscalar approach to exploring differences in voices (who) describing wild(er)ness (what). Here, the starting point was an analysis of individual documents to explore the influence of gender on descriptions of wild places, in particular through the use of pronouns (a classic how question). The locations of these descriptions (for the initial analysis around Rannoch Moor, and then more broadly concentrated in the UK) are not explicitly considered, but, just like the language of analysis (English here), are important caveats, since these empirical results are specific to the languages, locales, and corpora analysed.

Nonetheless, historical writing and literature influences how environments are perceived today. Karen Jones with Diana Maynard and Flurina Wartmann looked at historical travel writing to identify patterns in how a particular location, Loch Lomond (where) was described (what). Karen and her group took a similar hybrid approach to that argued for by Joanna, emphasising their exploratory approach by naming it 'forensic fishing'. Interestingly, this study is the only one which puts locations from text into their geographic context on topographic maps. Furthermore, the study hints at an underlying process in the writing of these texts – the process of place-making, in this case through the genre of travel guides – and the diversity of voices ignored in this writing.

Tobias Zuerrer takes a different approach to this forensic fishing in his analysis of *Frankenstein*, and starts from a guiding question: whether there are differences in how urban and natural landscapes are conceptualised in the novel. His question can be seen as an exploration of what and where, guided by a specific and well-formulated research question.

Our final study, and the only one to specifically develop new methods, is all about location, and as such poses the question as to where specific passages can be located – an important task if we are to georeference using more complex approaches than the simple toponym-lookup applied by Karen Jones and her team. Simon Scheider and colleagues did so by developing and implementing a model which captures different ways (a how question) spatial information is conveyed through frames of references.

Analysing our case studies, we see much more diversity in the questions posed than the methods used. We find ample examples of studies addressing five of the six questions (what, when, where, how and who) and it appears that the relative homogeneity of computational methods is overcome by on the one hand the diversity of the sources used by the teams, and on the other by the disciplinary backgrounds of those working on a particular problem. Each of these questions can, at least at a superficial level, be answered by, for example, counting and extracting words appropriately. Thus, for example, we could explore ways in which climate is described over time by extracting and comparing adjectives for some given corpus, stratified by time. Of course, as our case studies show, by combining micro and macroreadings of material, it is possible to interpret in much more depth. However answering (as opposed to posing) why questions computationally requires, we suggest, building upon the foundations laid in this book.

Since this book is about environmental narratives, it is also worth reflecting on the lenses through which our teams explored the environment. Change was an important theme – both as anthropologically driven climate change in the context of changing attitudes to glaciers, and with respect to attitudes as the urban environment of the Canal District was 'revitalized'. The influence of the past on current ways of exploring Loch Lomond and its environs can be seen as one way of exploring what and how a particular landscape is valued, contrasting effectively with the analysis of a particular contemporary management genre, Landscape Character Assessment. Potential differences between the ways in which particular sorts of environments are described in a single text, the urban and natural of Mary Shelley's *Frankenstein* suggests another way of exploring how the environment is implicitly and explicitly valued in text. Turning all of these ideas on their head, and looking not at the observed, but the observer, and studying the influence of gender on ways in which environments are written about reminds us that narrative should be studied in context, and that power (who writes, which languages are digitised, for which languages are tools available, who researches) has important consequences for any interpretations. Equally, understanding the diversity of ways that locations can be described, and recognising that this is not language-independent, emphasises the need to develop methods applicable not across cultures, but rather to cultures. Although we almost exclusively worked with English, it is also worth pointing out that the methods applied may struggle with texts not written in modern English amenable to use of basic text analysis.

In a book of this kind, it is traditional to close with a research agenda. Such agendas though often become prescriptive, limiting the diversity of research in a field and constraining imagination. With this in mind, we have chosen to present not a research agenda, but rather some starting points for future work, which we believe might be fruitful in developing the potential of computational analysis of environmental narratives. Our ideas are seeded by the potential demonstrated by the case studies in this book.

Perhaps one of the most surprising findings in hindsight was the power of computational analysis in individual volumes or small, curated corpora, rather than the big data analysis so often trumpeted as the way forward in contemporary research. Our teams and their interdisciplinary compositions were much better suited to a productive mixture of macro and micro-analysis. This microanalysis, was most productive where the thematic specialists were involved in the selection of texts and their qualitative exploration. At the scale of our case studies, which can be seen as pilots for future research, an important limiting factor was the volume of text which could be read, rather than computationally analysed. We suggest that future work on environmental narrative takes heed of this productive combination of well-chosen research questions and small, thematically focused collections as a starting point for research.

A second key finding of the work carried out in this book was the utility and effectiveness of existing, well-known methods for text analysis. Although the promise of machine learning and machine understanding of text is one which has gained much attention in natural language processing, we believe that effective research on environmental narratives should take advantage of existing, well-established and crucially, well-understood methods. For example, simple approaches to classification can deliver more than adequate performance for many of the questions of interest to our teams. Research on the computational analysis of environmental narratives should of course take advantage of general trends in text analysis. However, since the focus of work on environmental narratives is on explanation and understanding, future research should consider carefully how to combine existing, well understood methods with the most potential to generate insightful results.

This finding does not however preclude methodological development of particular relevance to environmental narratives. The study by Simon Scheider and colleagues on frames of references is an excellent example of such a study, since the ways in which locations are described in narratives of this nature are very rich, and unlikely to be identified in more general corpora such as news or social media. Working on a particular genre of texts was informative and led to much more productive research. This leads us to our third suggestion: that the development of methods can be productive when driven by use cases and collections directly relevant to understanding the environment. This emphasis on understanding, as opposed to methods in isolation, chimes with current debates in the digital humanities about the use of computational methods (Robertson and Mullen, 2021), and we believe it is an important result of the truly interdisciplinary process of writing this book.

Development of new methods presupposes that a variety of resources exist. These include collections of text about the environment, annotations of texts with respect to, for example, landscape preference or sentiment and resources such as environmentally specific gazetteers and lexicons. In writing this book all of these elements were hard to find, and we encourage future researchers to give much more consideration to not only reproducibility and replicability at the level of individual publications, but also considering shared tasks and resources more intensively to further research. Examples with great potential for more effective use include the Corpus of Lake District Writing (Rayson et al., 2017), the Text+Berg corpus (Volk et al., 2010) and resources available through APIs such as the Guardian's Country Diary, all of which were used within the case studies in this book.

As we wrote the text accompanying our case studies, one major challenge was exploring results, especially from larger datasets, with static visualisations, often limited to displaying the most highly ranked terms. More effective visualisations, capable of linking different views and moving beyond simply placing documents or texts on a map remains an important challenge for future work. Computational analysis of environmental narratives could thus be a very productive area for interdisciplinary work on effective and efficient visualisation and visual analytics. This need for exploration is important, as the approaches developed here will provide the bridge between macroanalysis and microreading. Providing more integrated ways to move from overviews of datasets, through zooming and filtering to details on demand, as proposed by Schneiderman in his influential information-seeking mantra (Shneiderman, 2003) would be an important contribution if we are to more effectively interpret material in context.

Productive interdisciplinary work lay at the heart of the case studies around which this book is based, and also lies at the core of our last recommendation. Future work should start not from the identification of methods or datasets, but with productive questions posed by experts with underlying thematic expertise. The potential of environmental narrative as source for computational analysis mediated by humans is, we believe enormous, and hope this book can stimulate future work in the field.

References

Rayson, Paul, Alex Reinhold, James Butler, Chris Donaldson, Ian Gregory, and Joanna Taylor (2017). "A deeply annotated testbed for geographical text analysis: The corpus of lake district writing". In: *Proceedings of the 1st ACM SIGSPATIAL Workshop on Geospatial Humanities*, pp. 9–15. DOI: 10.1145/ 3149858.3149865.

Robertson, Stephen and Lincoln Mullen (July 2021). "Arguing with digital history: Patterns of historical interpretation". In: *Journal of Social History* 54.4, pp. 1005–1022. ISSN: 1527-1897. DOI: 10.1093/jsh/shab015.

Shneiderman, Ben (2003). "The eyes have it: A task by data type taxonomy for information visualizations". In: *The craft of information visualization*. Amsterdam: Elsevier, pp. 364–371. DOI: 10.1016/B978-155860915-0/50046-9.

Volk, Martin, Noah Bubenhofer, Adrian Althaus, Maya Bangerter, Lenz Furrer, and Beni Ruef (2010). "Challenges in building a multilingual alpine heritage corpus". In: *Seventh International Conference on Language Resources and Evaluation (LREC)*.

www.ingramcontent.com/pod-product-compliance
Lightning Source LLC
LaVergne TN
LVHW011803070326
832902LV00026B/4618